TREES AND SHRUBS OF BRITISH COLUMBIA

*To Cam. Place,
with compliments of
Chris Brayshaw.*

ROYAL BRITISH COLUMBIA MUSEUM
HANDBOOK

TREES AND SHRUBS
OF BRITISH COLUMBIA

written and illustrated by

T. CHRISTOPHER BRAYSHAW

UBC PRESS / VANCOUVER

Published by UBC Press in collaboration with the Royal British Columbia Museum.

The Royal British Columbia Museum gratefully acknowledges Forest Renewal British Columbia (FRBC) and the Ministry of Forests for financial assistance in the production of this book.

Printed and bound in Canada by Friesens.
(See page 374 for photographic credits.)

Canadian Cataloguing in Publication Data

Brayshaw, T. Christopher, 1919-
   Trees and shrubs of British Columbia

   (Royal British Columbia Museum handbook, ISSN 1188-5114)

   Previous ed. published: The trees and shrubs of British Columbia / E.H. Garman. British Columbia Provincial Museum, 1973. (Handbook no. 31) Co-published by the Royal British Columbia Museum.
   Includes bibliographical references: p. 357
   ISBN 0-7748-0564-1

   1. Trees – British Columbia – Identification. 2. Shrubs – British Columbia – Identification. I. Garman, E.H. The trees and shrubs of British Columbia. II. Royal British Columbia Museum. III. Title. IV. Series.

QK203.B7B72 1996   582.16097111   C96-960205-7

UBC Press   Royal British Columbia Museum
University of British Columbia   675 Belleville Street
6344 Memorial Road   Victoria, British Columbia
Vancouver, British Columbia   V8V 1X4 Canada
V6T 1Z2 Canada

Phone: (604) 822-3259
Fax: 1-800-668-0821
E-mail: orders@ubcpress.ubc.ca

CONTENTS

PREFACE

This book is a direct descendant of the *Guide to the Trees and Shrubs of British Columbia* by Dr B.G. Griffith, Professor of Forestry at the University of British Columbia. Griffith's guide first appeared as a mimeograph in 1934, to assist students and woodsmen, and later as a printed publication, produced by the British Columbia Forest Service (now the Ministry of Forests) in 1937.

The second (1953), third (1963) and fourth (1970) editions were prepared by Eric H. Garman of the British Columbia Forest Service, and contained thoroughly researched revisions of the species descriptions and records of distribution.

The British Columbia Provincial Museum published the fifth edition in 1973 (22), and the guide joined the Provincial Museum Handbook series as Handbook 31. Few changes were made to the content of the guide at that time.

For the current work, the Ministry of Forests requested in 1991 an illustrated handbook based on the guide as it had existed until then. This was a much larger project than the earlier guides, and necessitated much more research and more intensive examination of specimens and records.

In general, I have retained the form and content established by Garman, but I have also made extensive changes and additions. Most of the text and keys have been rewritten to accommodate recent advances in our knowledge. In general, the changes fall into the following categories:

(1) *Additional species.* In the process of research for this work, many more woody species, both native and exotic, were found to be growing wild in British Columbia than had previously been known.

(2) *Taxonomic and nomenclatural changes.* Partly as a result of the recent reorganization of some genera, there are many changes in the names of species. Where these changes occur, the names used in former editions of the guide are indicated.

(3) *Revisions to keys.* Most of the keys have required extensive revision, in large part because of the need to accommodate the many species and genera and some families that have been added to the total species roll for this work. I have also composed a general key to the genera, based

as far as possible on vegetative characters. Hopefully this will make it easier to identify specimens in the absence of flowers or fruit.

(4) *References.* In preparing this work, I consulted many publications that were not available or not used when earlier guides were prepared. The list of references has therefore been reorganized and the entries renumbered.

Since this is a fully illustrated work, specific page references to *Native Trees of Canada* (29, quoted as *NT* in earlier editions) and to the *Forest Flora of Canada* (11, quoted as *FF* in earlier editions) seemed unnecessary and have been omitted. Nonetheless, *Native Trees of Canada,* now in its eighth edition, is a useful source of illustrations and other information.

Some of the material dealing with the catkin-bearing families, especially the willows, has been adapted from the revised edition of my book *Catkin-Bearing Plants of British Columbia* (5), which was published in April 1996.

(5) *Illustrations.* Preparing the illustrations occupied more than half the time it took to create this work. I have arranged the drawings of 314 species and subspecies into 76 plates. In most cases, similar species are illustrated side by side on the same plate, even when they are not closely related genetically. Individual drawings are numbered and provided with scale bars. Some of the illustrative material has been copied from an earlier publication (3) by this writer, with permission from the Canadian Forest Service.

Information arrives continuously, but there must come a moment of decision to cut off the flow so that publication can proceed. This work was essentially completed in 1994, although I have made some minor corrections and adjustments since then. Consequently, I did not use literature and records that became available after 1994.

I acknowledge with gratitude the staff of the Royal British Columbia Museum, especially Dr Robert T. Ogilvie, Curator of Botany (now retired), who supervised the work and helped in many ways, and John Pinder-Moss, Manager of the Herbarium, who made specimens available for study and illustration. Acknowledgement is also due to Eric H. Garman, who provided new information in an annotated copy of his guide and a file of notes.

INTRODUCTION

Trees and Shrubs

It is important at the outset to define the group to be dealt with in this work, since decisions on the choice of species for inclusion depend on one's concept of what is a tree or a shrub. These are subjectively viewed groups that in nature are not sharply distinguished from each other or, as a group, from herbaceous plants. To aid in making such decisions, some arbitrary criteria must be accepted to circumscribe each group.

To qualify for inclusion in this book, a plant must have a perennial, more or less woody stem above ground that bears overwintering buds and does not die back to a basal stem or rhizome in the soil in winter. This group includes trees, shrubs and woody vines. Trees have a single stem from the ground up and are at least 2 metres tall. Shrubs are multi-stemmed from the ground and can be any size. Vines are so slender that, whether twining or not, they cannot support their own weight without the help of other vegetation.

Notwithstanding the above definitions of trees and shrubs, some ambiguities occur. Several woody species grow as trees under favourable conditions, but as shrubs in harsher environments. For this reason, some genera appear under both headings in the general key to the genera. Another aspect of this question is displayed by the Aspen Poplar (*Populus tremuloides*), which we think of as a tree without question, and the Sandbar Willows (*Salix exigua* and *S. sessilifolia*), which we see as typical shrubs. All these plants have the same habit of growth, with extensive shallow root systems giving rise at intervals to erect shoots of more or less treelike form above ground. Here we fall back on size as the distinguishing criterion. Otherwise, the Aspen and all our native Cherries would technically be considered shrubs.

At the small end of the scale, the distinction between dwarf shrubs, creeping shrubs and herbs may be drawn rather arbitrarily, according to one's judgement. Nature knows nothing about our vocabulary. I accepted as a shrub any plant that regularly maintains overwintering, living stems and buds above ground level.

Classification and Nomenclature

Biologists classify all forms of life into a hierarchy of groups or categories of progressively lower rank, narrower scope and smaller size. Each group is an aggregate of one or more groups of the next lower rank. (7)

The classification of the living world, and in particular of the Plant Kingdom, has, during the twentieth century, undergone several major revisions, especially in the higher categories. While the families and lower categories of the flowering plants have suffered relatively minor alterations, their arrangement in the orders and higher categories has undergone many adjustments in adaptation to the rapid development of our understanding of the genetic relationships among the major groups of flowering plants.

In accordance with the recent system of classification proposed by Arthur Cronquist (11), the flowering plants are classified into the following categories (with examples of plants covered in this book):

Division; e.g., Magnoliophyta (the Angiosperms, or flowering plants)
Class; e.g., Magnoliopsida (the Dicotyledons)
Subclass; e.g., Dilleniidae
Order; e.g., Salicales (Willow Order)
Family; e.g., Salicaceae (Willow Family)
Genus; e.g., *Salix* (the willows)
Species (sp.); e.g., *reticulata* (Netted Willow)
Subspecies (subsp.); e.g., *reticulata*
Variety (var.); e.g., *reticulata*
Forma (f.); e.g., *villosa.*

Intermediate levels of classification, such as subfamily, tribe, subgenus and series are sometimes used if they are required, but in categories below the level of species, they are not essential. Other classification systems have been proposed, but they differ mainly in their handling of the major categories in the hierarchy, leaving family, genus and species unaffected.

The technical (scientific) name used to identify a plant species is a combination of the name of the genus to which it belongs and that of the species itself. This is known as the Binomial System of Nomenclature. The technical names are in Latin, a convention that has the following advantages:

(1) The Latin system is international and has worldwide application. While vernacular "common" names for plants vary from region to region and with the language spoken, Latin names are standard everywhere, regardless of the local language.

(2) Since Latin is no longer in popular use, the meanings of its words are constant, while the meanings of words in languages still in everyday use are changing with time as well as varying regionally from dialect to dialect.

(3) Many inconspicuous plants have no common names, but every species found is given a Latin name when its description is published. Original descriptions of newly discovered species are published in Latin for the above reasons.

(4) This binomial system of Latin names reflects the true relationship among species. If species are closely enough related to be included in the same genus, then they carry the name of that genus as part of their complete names. Common names can be misleading. For example, the plant often called Wolf-willow is neither a true willow nor even closely related to willows. Wolf-willow's Latin name is *Elaeagnus commutata*. True willows belong to the genus *Salix*.

The genus name is a noun and the species name is an adjective modifying the noun: *Salix* is the classical name for the willows, and *reticulata* means "netted". Thus the specific name, *Salix reticulata*, refers to a species of willow recognized by the conspicuous netted pattern of veins on its leaves. Subspecies names are also adjectives (as are those for varieties and forms). Subspecific names are preceded by an abbreviation indicating its rank within the species. A subspecific group that includes the typical example of the species carries a name that repeats that of the species. For example, *Salix reticulata* subsp. *reticulata* is the typical form of the species. When discussing named species in a genus, one can avoid repeating the generic name by using only its initial letter after the first mention (e.g., *Salix reticulata* and *S. arctica* belong to the same genus).

Because botanists often disagree about the classification and naming of plants, it is customary to identify the author who was responsible for describing and naming the plant. This helps to avoid ambiguity in the application of names. The author's name or abbreviation is written after the name of the species, such as "*Salix exigua* Nuttall" and "*Salix interior* Rowlee".

If the position of a plant in the classification system is revised, the original author's name is placed in parentheses and followed by the name of the person who has reclassified it. For example, Arthur Cronquist reallocated *Salix interior* to a subspecies of *S. exigua,* stating that the population named "*interior*" by Rowlee is not a distinct species, but falls within the circumscription of *S. exigua* as described by Nuttall. Cronquist's reclassification is indicated by the citation "*Salix exigua* Nuttall subsp. *interior* (Rowlee) Cronquist."

It is not uncommon for a description of a species provided by one author to be published in a book written by another author. In this case, the names of both authors may be cited and linked with the terms *"ex"* or *"in"*. Thus, the citation of *Quercus garryana* Douglas *ex* Hooker indicates that Hooker (in *Flora Boreali Americana*) published a description that was provided by Douglas.

It is often found that two or more authors have given names to the same species. When this occurs, the name to be published first, normally, is the correct one to use. However, this apparently simple principle is complicated by a system of rules that affects its application and makes decisions more difficult in some cases (25). When more than one name is in use for a species, the author of a book such as this may indicate them all: first, he uses the name he considers to be the correct one; then he lists any other names that have been applied to the same plant, usually in parentheses. These other names are called "synonyms". Synonyms are useful references to publications by authors who may record the same plant under a different name, which they believe is the correct one.

Hybrids

As a general rule, cross-breeding does not occur between individuals of different species. Normally, the character differences that distinguish species are invariable, and species are readily distinguishable from each other. On the other hand, individuals of different subspecies or varieties can interbreed freely, so they naturally tend to merge into one another in character through a series of intermediate forms. (7)

Among some groups of plants, however, the genetic barriers to cross-breeding between species are not always completely effective. As a result, crossing sometimes occurs between related species, and hybrids are produced. A hybrids can have characteristics that are intermediate between those of the parent species or a combination of characteristics of both species.

Hybrids, when fertile, may cross with their parent species or with each other. In this way intergrading series of forms are sometimes produced that blur what are normally clear distinctions in character between the species. A populations of such mixed origin is called a "hybrid swarm". Hybrid swarms make it difficult to identify the species found among them.

Once a plant has been identified as a hybrid, it is named using a formula that includes the names of the parent species connected by a multiplication sign (×); e.g., *Betula occidentalis* × *papyrifera*. Many hybrids have been named as separate species before their hybrid origins were realized. When a plant once thought to be a species is later determined to be a hybrid, its mixed origin indicated by inserting the "×" between the generic and specific names. *Betula* × *piperi* Britton is the recognized name of the hybrid between *B. occidentalis* and *B. papyrifera*, Britton having described *B. piperi* as a species before its hybrid origin was understood.

The Organization of This Book

The Order of Families

As in earlier guides and in many other botanical manuals, the order of the plant families in this book follows, in general, the Englerian order. This order became so well established and so well learned by previous generations of botanists, that, although it is no longer regarded as representing the most natural relationships among the families, most botanists can find their way about in it without difficulty. However, within the families, the order of the descriptions of the genera and species in this work is determined largely by my effort to juxtapose the illustrations of closely related or similar species. This order is reflected in the checklist of species.

Botanical Language

To those coming to botanical writing for the first time, the plant descriptions, with their verbless sentences and technical shorthand, may be a little daunting. Some patience and practice with the descriptions and keys should cure any apprehensions. To be succint, I have followed the botanical practise of omitting unnecessary words. Many verbs and articles are implied. Other words and phrases are short forms commonly in use by botanists: *ours*, as in "bracts small in ours", refers to British Columbian species; the meanings of "longer than wide" and "as long as wide" are clear enough and shorter than the grammatically correct equivalents; the statement "leaflets (5-)7-9(-13) means that the leaf usually has 7 to 9 leaflets, but rarely as few as 5 or as many as 13. Botanical terms, which are more precise and succinct than everyday English, and words with specific botanical meanings, are defined in the illustrated glossary at the back of the book.

The Working Book

The working body of this book is introduced by a series of diagnostic keys, through which a user can make a succession of choices between alternative characteristics, choosing the better fit at each successive choice and eventually arriving at a name for the plant at hand. It is always advisable to check the diagnosis arrived at through the keys against the full description of the species in the text.

There are keys to the families, one to native trees in winter and one directly to the genera. Although very long, this last key is probably the most useful for general application. Under each family description is a key to the genera in the family if there is more than one; and under each genus is a key to its species, if needed. There are thus two alternative routes through the keys to the identity of a plant specimen.

In these indented keys, the alternatives at each stage are indented by the same amount so that the second alternative starts vertically below the first. All the choices in the keys are numbered, and in any situation where a choice has to be made, the alternatives are indicated by "a" and "b" (and rarely "c").

I have tried to keep the illustrations of the species as close as possible to their descriptions. Also, the description of each species is followed by a reference to the plate and figure where the species is illustrated.

A short supplement contains 96 colour photographs illustrating 92 of the species in this book. These photographs are cited in the text where they are relevant.

References to Other Works

I have consulted many other publications during the preparation of this book. They appear in a numbered list under References at the back of the book and are cited in the text by their numbers in parentheses. These publications may be of interest, because they present more detailed discussion of the information summarized here.

Distributions of Species

In this work, the ranges of species are stated in general terms, without the use of range maps. Specific localities are not given unless they represent known limits of distribution or occurrences of rare species.

More detailed information, mainly in the form of range maps, is given for British Columbia in publications by the Royal British Columbia Museum (formerly the British Columbia Provincial Museum) and the British Columbia Ministry of Forests (formerly the British Columbia Forest Service) (4, 5, 7, 54, 56, 57, 58, 59, 60). For North America in general, the *Atlas of United States Trees* (39, 40) presents distribution maps of all North American trees and some shrubs. For Alaska and other northern regions, maps and other information are contained in works by Hulten (30, 31), by Viereck and Little (61), and by Argus (1) specifically for the willows. Names of forest regions mentioned in range statements for species refer to the regions described by Rowe (53).

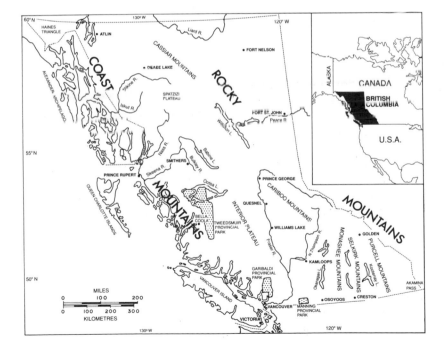

CHECKLIST

This checklist follows the same order as the species descriptions. Species listed after the first in a genus are indented without the genus name, and subspecies, etc., are indented further. The numbers on the right refer to the plate or plates where the plants are illustrated.

Pinaceae
Picea sitchensis (Bongard) Carriere 1, 3
   glauca (Moench) Voss 1, 3
      subsp. *glauca*
         var. *glauca*
         var. *albertiana* (Stewardson-Brown) Sargent
         var. *porsildii* Raup
      subsp. *engelmannii* (Parry *ex* Engelmann) T. Taylor
      × *lutzii* Little (= *P. glauca* × *sitchensis*)
   mariana (Miller) B.S.P. 1, 3
Tsuga heterophylla (Rafinesque) Sargent 1, 3
   mertensiana (Bongard) Carriere 1, 3
Pseudotsuga menziesii (Mirbel) Franco 2, 4
         var. *menziesii*
         var. *glauca* (Beissner) Franco
Abies grandis (Douglas *ex* D. Don in Lambert) Lindley 2, 4
   amabilis (Douglas *ex* Loudon) Forbes 2, 4
   balsamea (L.) Miller subsp. *lasiocarpa* (Hooker) Boivin 2, 4
Larix laricina (DuRoi) K. Koch 2, 5
   lyallii Parlatore in DC. 2, 5
   occidentalis Nuttall 2, 5
Pinus banksiana Lambert 1, 6, 8
   contorta Douglas *ex* Loudon 1, 6, 8
         var. *contorta*
         var. *latifolia* Engelmann *ex* S. Watson
      × *murraybanksiana* Righter & Stockwell (= *P. banksiana* × *contorta*)
   ponderosa Douglas *ex* Lawson 1, 6, 8
   monticola Douglas *ex* D. Don in Lambert 1, 7, 8

Polygonaceae

Ranunculaceae

Berberidaceae

Grossulariaceae

Hydrangeaceae

Rosaceae

Oleaceae

Loganiaceae

Polemoniaceae

Labiatae (Lamiaceae)

Solanaceae

Scrophulariaceae

Caprifoliaceae

Compositae (Asteraceae)

DIAGNOSTIC KEYS

The Divisions of the Seed Plants

The seed-bearing plants are divided into two Divisions or Phyla: Pinophyta (the Gymnosperms) and Magnoliophyta (the Angiosperms or flowering plants). In the Gymnosperms, which includes the conifers, the ovules and seeds are attached to scales arranged in cones or, in the Yew, attached to the apex of a vestigial cone and completely exposed; in the Junipers the seeds are embedded in the ripe cone by the overgrowing, fleshy cone scales. In the Angiosperms, which includes our broad-leaved trees and shrubs, the ovules and seeds are completely enclosed within a chamber or chambers (the ovary) in the centre of a true flower.

Key to the Divisions of Trees and Shrubs of British Columbia

Trees and shrubs with linear, needlelike or scalelike, usually evergreen leaves; stems all woody; fruit a woody or fleshy cone formed of seed-bearing scales, or an exposed seed in a fleshy cup
.... **Pinophyta (Gymnosperms)**
Trees and shrubs with usually broad, deciduous leaves; stems woody or partly herbaceous or succulent; fruit of diverse forms
.... **Magnoliophyta (Angiosperms)**

All the British Columbian Gymnosperms belong in the Class Pinopsida (the Conifers) and all our Angiosperm trees and shrubs are members of the Class Magnoliopsida (the Dicotyledons).

Key to the Families of Conifers (Class Pinopsida) in British Columbia

1a Leaves all needlelike and alternate, or in alternately arranged bundles; i.e., attached to the twig in a spiral arrangement, although sometimes twisted so as to project horizontally.

    2a Leaves flattened in cross section, abruptly pointed, never in bundles, the petiole base extending downward on the twig. Fruit an exposed seed sitting in a red fleshy cup (aril) . . . . **Taxaceae**

    2b Leaves either not flattened in cross section or not abruptly pointed, or jointed to a rounded leaf scar; attached singly or forming bundles. Fruit a dry cone with scales overlapping the many, usually winged, seeds . . . . **Pinaceae**

1b Leaves and cone scales opposite or whorled. Leaves mostly scalelike and overlapping. Cones sometimes berrylike . . . . **Cupressaceae**

Key to the Families of Native or Naturalized Woody Angiosperms (Class Magnoliopsida) in British Columbia

1a Stems succulent, green, almost leafless, spiny . . . . **Cactaceae**

1b Stems not succulent.

    2a Petals wanting, sepals present or absent.

        3a Flowers minute, unisexual, without petals, in catkins or similar condensed clusters.

            4a Leaves pinnately compound.

                5a Leaves alternate . . . . **Juglandaceae**

                5b Leaves opposite.

                    6a Leaflet margins shallowly lobed . . . . **Aceraceae**

                    6b Leaflet margins unlobed . . . . **Oleaceae**

            4b Leaves simple.

                7a Staminate flowers only in catkins; fruit an acorn; leaves deeply pinnately lobed . . . . **Fagaceae**

                7b Both staminate and pistillate flowers in catkins; fruit various, not an acorn; leaves usually unlobed.

                    8a Seed with tuft of hairs; fruit a capsule . . . . **Salicaceae**

                    8b Seed without a tuft of hairs; fruit one-seeded and not dehiscent.

                        9a Leaves aromatic, dotted, oblanceolate, acute-tipped, toothed only near tips . . . . **Myricaceae**

                        9b Leaves not aromatic, ovate or orbicular; acute to rounded at tips, toothed all around.

10a   Leaves pinnately veined; fruit dry: nuts, nutlets or samaras (winged nutlets) . . . . **Betulaceae**
10b   Leaves strongly 3-veined from base, sometimes lobed;
fruit fleshy, blackberry-like . . . . **Moraceae**
3b   Flowers not in catkins.
11a   Trees; fruit winged, dry.
12a   Leaves alternate, simple; fruit winged all around
. . . . **Ulmaceae**
12b   Leaves opposite, compound; fruit with long terminal
wing . . . . **Oleaceae**
11b   Shrubs, vines or small trees; fruit not winged.
13a   Fruit dry nutlets.
14a   One 3-angled, tailless nutlet per flower. Leaves alternate, simple. Low shrub . . . . **Polygonaceae**
14b   Clustered 2-edged nutlets with long feathery "tails."
Leaves opposite, compound. Vine . . . . . **Ranunculaceae**
13b   Fruit berrylike.
15a   Leaves scurfy . . . . **Elaeagnaceae**
15b   Leaves not scurfy.
16a   Leaves linear, 1 cm or less long . . . **Empetraceae**
16b   Leaves broader and longer.
17a   Leaves narrowly oblanceolate and obscurely
veined . . . . **Thymelaeaceae**
17b   Leaves elliptic, orbicular or broadly oblanceolate, and distinctly veined . . . . **Rhamnaceae**
2b   Petals and sepals both present.
18a   Petals separate.
19a   Fruit dry, or at most the receptacle fleshy.
20a   Fruit of separate carpels or a single carpel.
21a   Flower radially symmetrical . . . . **Rosaceae**
21b   Flower bilaterally symmetrical
. . . . **Leguminosae (Fabaceae)**
20b   Fruit of united carpels.
22a   Leaves opposite.
23a   Leaves compound.
24a   Leaves palmately compound
. . . . **Hippocastanaceae**
24b   Leaves pinnately compound . . . . . . **Aceraceae**
23b   Leaves simple.
25a   Leaves entire or simply toothed.
26a   Flowers large, white . . . . . **Hydrangeaceae**

       26b   Flowers small, reddish to purplish
       . . . . **Celastraceae**
      25b   Leaves palmately lobed . . . . **Aceraceae**
  22b Leaves alternate or absent.
     27a Leaves strongly 3-veined from base
     . . . . **Rhamnaceae**
     27b Leaves pinnately veined . . . . **Ericaceae**
19b Fruit more or less fleshy, a berry, pome or drupe.
  28a Leaves compound.
     29a Leaves evergreen, spiny, with stipules . . **Berberidaceae**
     29b Leaves deciduous, not spiny.
       30a Leaves without stipules . . . . **Anacardiaceae**
       30b Leaves with stipules . . . . **Rosaceae**
  28b Leaves simple.
     31a Leaves opposite . . . . **Cornaceae**
     31b Leaves alternate.
       32a Winter buds naked . . . . **Rhamnaceae**
       32b Winter buds enclosed by scales.
         33a Leaves prickly.
           34a Leaves large, palmately veined, deciduous;
           stem prickly . . . . **Araliaceae**
           34b Leaves small, pinnately veined, evergreen; stem
           not prickly . . . . **Aquifoliaceae**
         33b Leaves not prickly.
           35a Leaves with persistent stipules . . . . . **Rosaceae**
           35b Leaves without stipules, or stipules early decidu-
           ous.
             36a Leaves palmately lobed . . **Grossulariaceae**
             36b Leaves not lobed.
               37a Branches bearing spinelike leaves at
               nodes, with normal leaves in the axils of the
               spines . . . . **Berberidaceae**
               37b All leaves of normal type . . . . **Rosaceae**
18b Petals united.
  38a Ovary superior.
     39a Flowers bilaterally symmetrical.
       40a Corolla tubular, 2-lipped at mouth.
         41a Leaves aromatic when crushed. Stem square in sec-
         tion. Fruit of 4 nutlets . . . . **Labiatae (Lamiaceae)**
         41b Leaves not aromatic. Stem not square in section.
         Fruit a capsule . . . . **Scrophulariaceae**

40b Corolla bell-shaped, 5-lobed, with one lobe often
coloured differently . . . . **Ericaceae**
39b Flowers radially symmetrical.
42a Corolla bowl-shaped, with 10 pouches . . . . . . **Ericaceae**
42b Corolla tubular to urn-shaped.
43a Corolla urn-shaped, with a narrow mouth
. . . . **Ericaceae**
43b Corolla tubular or funnel-like, with spreading lobes.
44a Corolla lobes 4; leaves opposite, wide.
45a Leaves densely felted beneath; flowers purple
. . . . **Loganiaceae**
45b Leaves glabrous; flowers white . . . . . **Oleaceae**
44b Corolla lobes 5; leaves alternate or, if opposite,
linear.
46a Fruit a berry . . . . **Solanaceae**
46b Fruit a capsule . . . . **Polemoniaceae**
38b Ovary inferior.
47a Flowers small, in dense heads; fruit a dry achene
. . . . **Compositae (Asteraceae)**
47b Flowers not in heads; fruit berrylike.
48a Leaves alternate . . . . **Ericaceae**
48b Leaves opposite . . . . **Caprifoliaceae**

**Key to Winter Identification of Deciduous Angiosperm Trees
Native to British Columbia**

1a Leaf scars opposite.
2a Leaf scars meet around twig; pith small.
3a Terminal buds present.
4a Terminal buds with 2 long, narrow scales covering entire
bud; lateral buds minute . . . . *Cornus*
4b Terminal buds with usually more than 2 scales visible;
lateral buds prominent.
5a Terminal buds with 3-4 pairs of overlapping scales;
bundle scars 5-9; twigs coarse . . . . *Acer macrophyllum*
5b Terminal buds with 1-2 pairs of outer scales.
6a Twigs slender, green to red; leaf scar with 3 bundle
scars; buds same colour as twig, glabrous
. . . . *Acer glabrum*
6b Twigs stout, grey; leaf scar with several bundle scars in
arc; buds brown, hairy . . . . *Fraxinus latifolia*

3b   Terminal buds absent; bundle-scars in threes
. . . . *Acer circinatum*
2b   Leaf scars do not meet around twig; winder-buds with many acute scales; pith very large . . . . *Sambucus*
1b Leaf scars alternate.
   7a   Bud scales 0-3.
      8a   Buds naked, without bud scales . . . . *Rhamnus*
      8b   Buds enclosed by 1-3 scales.
         9a   Lateral buds with 1 scale, no terminal bud; pith circular
. . . . *Salix*
         9b   Lateral buds with 2-3 scales, terminal bud present; pith 3-angled in cross section . . . . *Alnus*
   7b   Bud scales more than 3.
     10a   Buds clustered at tips of twigs.
        11a   Pith 5-angled in cross section . . . . *Quercus*
        11b   Pith circular in cross section . . . . *Prunus*
     10b   Buds single at tips of twigs.
        12a   Axillary bud with lowest bud scale directly above leaf scar. Pith 5-angled in cross section . . . . *Populus*
        12b   Axillary bud with lowest scale not directly above leaf scar. Pith not 5-angled.
          13a   Bud tips acute; branchlets without spines or stout spurs.
            14a   Buds grey-brown; bark with horizontally elongated lenticels . . . . *Betula*
            14b   Buds reddish; bark with minute lenticels not horizontally elongated . . . . *Amelanchier*
          13b   Bud tips obtuse or rounded; branchlets stoutly spurred or spiny.
            15a   Short branchlets forming stout, hairy spurs with terminal buds . . . . *Malus*
            15b   Some short branchlets forming sharp, budless, glabrous spines . . . . *Crataegus*

Key to the Genera of Trees and Shrubs of British Columbia, Based Primarily on Vegetative Characters

1a   Stems succulent .... *Opuntia*
1b   Stems not succulent.
   2a   Leaves linear, needlelike or scalelike.
      3a   Trees.
         4a   Leaves scalelike.
            5a   Ultimate twig and leaves flattened, about 2 mm wide. Branches forming fernlike sprays that are smooth-feeling. Cones erect, ellipsoidal with overlapping scales. Old trees buttressed at bases .... *Thuja*
            5b   Ultimate twig, with leaves, cylindrical or 4-angled. Cone globose. Branches not fernlike, or if so, rough-feeling.
               6a   Ultimate twig and leaves cordlike, about 1 mm thick, cylindrical. Each leaf with a tiny spot (a resin vessel) on the back. Ovulate cone berrylike .... *Juniperus*
               6b   Ultimate twig and leaves 4-angled, the leaves distinctly keeled and sharp-pointed, unspotted on back; the branches forming flattish, hanging sprays that are rough to feel. Ovulate cone dry with centrally attached non-overlapping scales .... *Chamaecyparis*
         4b   Leaves needlelike.
            7a   Leaves in tufts, at least on older twigs.
               8a   Leaves evergreen, in tufts of 2-5 .... *Pinus*
               8b   Leaves deciduous, soft, in tufts of 10 or more on older twigs, separate on current year's twigs .... *Larix*
            7b   Leaves all separate, not tufted.
               9a   Leaves more or less 4-angled, on prominent projections on the twigs .... *Picea*
               9b   Leaves flattened; the base attachment either not or scarcely projecting from twig surface.
                  10a   Buds encased in resin. Cones erect, disintegrating before falling .... *Abies*
                  10b   Buds not encased in resin. Cones not erect, not disintegrating.
                     11a   Terminal buds conspicuous, pointed. Leading shoot erect. Cone hanging with projecting 3-pronged bracts .... *Pseudotsuga*
                     11b   Terminal bud small, inconspicuous, blunt.

12a Leading shoot inclined or drooping over.
Twigs pale yellowish brown. Leaf tips blunt.
Cone declining, dry .... *Tsuga*
12b Leading shoot erect. Twigs green. Leaf tips
sharply mucronate. Fruit a seed in a red fleshy
cup (aril) .... *Taxus*
3b Shrubs.
13a All leaves linear, often needlelike.
14a Leaves opposite or in whorls.
15a Leaves opposite, well separated; stem very pale
.... *Phlox*
15b Leaves whorled, although sometimes irregularly so.
16a Leaves in whorls of 3, very sharp-pointed and
stiff; stems ascending to horizontal, stout .. *Juniperus*
16b Leaves in often irregular whorls of 4, softish and
blunt; stems slender, trailing .... *Empetrum*
14b Leaves not whorled or opposite, although often densely
crowded.
17a Leaves in lateral tufts of 5, stiff and firm
.... *Pinus* (**krummholz form**)
17b Leaves alternate and separate, or if in lateral tufts,
soft and pale.
18a Leaves soft, hairy, light green or greyish; flowers
yellow, small, in terminal clusters of small many-
bracted heads. Low, broomy shrubs of semi-arid areas.
19a Flower heads with 4 bracts; twigs pale yellowish brown under white down .... *Tetradymia*
19b Flower heads with at least 10 bracts; twigs pale
grey-green or brownish.
20a Flower heads with bracts overlapping in 4 or
5 vertical rows; no ray flowers
.... *Chrysothamnus*
20b Flower heads with bracts not in vertical
rows; ray flowers present .... *Haplopappus*
18b Leaves firmer, stiff or leathery, glabrous or somewhat glandular, deep green.
21a Leaves 4-angled in cross-section, attached to
projecting bases on twigs
.... **Picea** (**krummholz form**)
21b Leaves not 4-angled. Leaf bases not projecting.

22a Buds encased in resin . . . . *Abies* **(krummholz form)**
22b Buds not encased in resin.
   23a Big shrub with young twigs green and mucronate leaves
   spread in a horizontal plane
   . . . . *Taxus*
   23b Dwarf or trailing, flowering shrubs with ascending
   branches; twigs not green.
      24a Leaves 5 mm or less long. Slender trailing shrub with
      ascending branches
      . . . . *Harrimanella*
      24b Leaves 8-15 mm long. Compact shrub with ascend-
      ing, relatively stout stems . . . . *Phyllodoce*
13b At least some leaves scalelike, opposite.
   25a Foliage greyish green, hard, a resin vessel visible as a dark
   spot on the back of each leaf. Robust shrubs with ascending or
   prostrate branches . . . . *Juniperus*
   25b Foliage deep yellowish green, without resin vessels. Flowering
   shrubs with erect 4-angled branchlets.
      26a Dwarf, compact shrubs (up to 20 cm) with erect branchlets
      2-7 mm thick including the leaves, which lack or have only
      very short, basal lobes . . . . *Cassiope*
      26b Taller shrubs (up to 1 m) with branchlets appearing
      whorled and 1.5 mm or less thick including the leaves, each of
      which has a pair of prominent, acuminate basal lobes
      . . . . *Calluna*
2b Leaves with widened blades (may be dissected with linear lobes).
   27a Leaves whorled . . . . *Eriogonum*
   27b Leaves not whorled.
      28a At least some of the leaves opposite.
      29a Leaves compound.
         30a Leaves palmately compound . . . . *Aesculus*
         30b Leaves pinnately compound.
            31a Climbing or trailing vine . . . . *Clematis*
            31b Upright shrubs or trees.
               32a Twigs grey, or greenish becoming grey; pith slen-
               der; inflorescence lateral.
                  33a Leaflets 3-5(-7), shallowly lobed; twig and
                  buds pale green and glabrous . . . . . . *Acer negundo*
                  33b Leaflets 5-11, entire or serrate; twigs grey;
                  buds brown or black, and hairy or glandular
                  . . . . *Fraxinus*

        32b Twigs yellowish green, becoming brown; pith
        relatively thick; inflorescence terminal; leaflets ser-
        rate, unlobed .... *Sambucus*
29b Leaves simple.
  34a Leaves palmately veined and lobed.
    35a Leaves sessile, small (1 cm or less long), deeply divided
    into needlelike, sharp-pointed lobes. Small spindly shrub of
    dry grassland .... *Leptodactylon*
    35b Leaves on petioles, larger (3 cm or more long).
      36a Twigs becoming pale grey. Buds with one pair of
      coherent scales. Shrubs .... *Viburnum*
      36b Twigs green or reddish, becoming dark red or brown.
      Buds with 2 or more pairs of scales. Trees or tall shrubs
      .... *Acer*
  34b Leaves pinnately veined.
    37a Buds naked, without bud scales.
      38a Leaves and twigs scurfy, with minute brown scales
      .... *Shepherdia*
      38b Leaves and twigs not scurfy .... *Rhamnus*
    37b Buds enclosed in bud scales.
      39a Leaves toothed, with small teeth.
        40a Leaves evergreen.
          41a Stems erect or ascending straight from the
          ground. Leaf pairs equally spaced. Flowers minute,
          red, sessile in leaf axils .... *Paxistima*
          41b Stems prostrate, or at least prostrate-based and
          then ascending. Flowers larger.
            42a Main stems elongate and prostrate. Leaf
            pairs equally spaced.
              43a Foliage aromatic when crushed. Flowers
              axillary, single. Corolla 2-lipped, usually
              white .... *Satureja*
              43b Foliage not aromatic. Flowers in pairs on
              erect peduncles that terminate short lateral
              branches. Corolla equally 5-lobed, usually
              pink .... *Linnaea*
            42b Stems prostrate-based and then ascending,
            not elongate. Leaf pairs concentrated toward end
            of year's growth. Flowers in terminal racemes,
            bluish purple, horizontally held, tubular with the
            mouth distinctly 2-lipped; the upper lip of 2
            lobes, the lower lip of 3 lobes ...... *Penstemon*

40b Leaves deciduous.
  44a Stems and undersides of leaves white-hairy. Flowers
    blue-purple, in long terminal panicles .... *Buddleja*
  44b Stems and leaves glabrous or almost so.
    45a Leaves finely and regularly serrate. Flowers reddish,
    5-petalled, stalked from leaf axils .... *Euonymus*
    45b Leaves with few scattered teeth. Flowers white, 4-
    petalled, in short, showy racemes on ends of lateral
    branchlets .... *Philadelphus*
39b Leaves entire or sinuately lobed.
  46a Leaves evergreen.
    47a Climbing vine .... *Lonicera*
    47b Erect or sprawling shrub.
      48a Leaves over 1 cm long, glaucous and glabrous beneath.
      Upright, although often small; shrub .... *Kalmia*
      48b Leaves not over 8 mm long, not glaucous, but finely pubes-
      cent beneath. Prostrate, mat-forming or cushion-forming shrub
      .... *Loiseleuria*
  46b Leaves deciduous.
    49a Buds attenuate-pointed, 4-10 mm long. Lateral leaf veins
    strongly arcuate, becoming parallel to leaf margin. Twigs 2-4 mm
    thick .... *Cornus*
    49b Buds acute to blunt, less than 5 mm long; veins not becoming
    parallel to margin.
      50a Lateral leaf veins running directly to margin; twigs very
      slender, commonly 1 mm or less thick ...... *Symphoricarpos*
      50b Lateral leaf veins arcuate, not running directly to margin.
        51a Vines with ovate leaves and terminal inflorescences
        .... *Lonicera*
        51b Shrubs.
          52a Leaves less than 5 cm long by 2 cm wide; inflores-
          cence a terminal panicle of white flowers .... *Ligustrum*
          52b Leaves larger: longer, wider or both; inflorescence an
          axillary pair of yellow flowers .... *Lonicera*
28b Leaves alternate.
  53a Leaves compound.
    54a Leaves absent, or present briefly and small, when at least some
    trifoliolate; branchlets deep green, slender, ascending nearly paral-
    lel, angled in cross section .... *Cytisus*
    54b Leaves regularly present throughout growing season; branch-
    lets variously coloured, generally not green.
      55a Leaves palmately compound or trifoliolate.

56a   Leaves palmate with more than 5 leaflets . . . . *Lupinus*
56b   Leaves with 3 or 5 leaflets.
  57a   Plant not prickly; leaves trifoliolate.
    58a   Leaflets with sinuate margins; fruit white berries in
    erect axillary racemes; shrubs. Poisonous to handle
    . . . . *Rhus*
    58b   Leaflets entire; fruit beanlike pods in hanging
    racemes; tree or shrub . . . . *Laburnum*
  57b   Plant bristly or prickly; leaves with 3 or 5 leaflets
  . . . . *Rubus*
55b   Leaves pinnately compound.
  59a   Stems bristly or prickly.
    60a   Leaflets entire. Stems with paired thorns by leaf bases.
    Buds sunk in twig. Suckering tree . . . . *Robinia*
    60b   Leaflets toothed along margins. Stem with usually scat-
    tered prickles, with or without paired thorns by leaf bases.
    Buds visible.
      61a   Stem with pinnate leaves its first (non-flowering) year
      . and trifoliolate leaves in later years . . . . *Rubus*
      61b   Stem living many years. Leaves always pinnate
      . . . . *Rosa*
  59b   Stems unarmed; glabrous or soft-hairy.
    62a   Leaves evergreen, hard, stiff, sharply toothed
    . . . . *Berberis*
    62b   Leaves relatively soft and thin.
      63a   Stipules present, at least at younger leaf bases.
        64a   Leaves with 5-7 leaflets. Prostrate or ascending
        shrubs to 1 m tall . . . . *Potentilla*
        64b   Leaves with 7-15 leaflets. Erect shrubs or trees to
        8 m tall . . . . *Sorbus*
      63b   Stipules absent.
        65a   Low compact grey-leafed shrubs with leaves less
        than 8 cm long . . . . *Artemisia*
        65b   Coarser, taller shrubs or trees, with green leaves
        more than 10 cm long.
          66a   Tree with evident buds in leaf axils. Twigs and
          leaves hairy and glandular. Fruit large drupes in
          hanging catkins . . . . *Juglans*
          66b   Colonial shrubs with buds concealed by leaf
          bases. Fruit small, in dense terminal panicles
          . . . . *Rhus*
53b   Leaves simple.

67a Vines; climbing trees or scrambling over adjacent bushes.
   68a Leaves evergreen, palmately veined. Inflorescence a terminal
     raceme of umbels . . . . *Hedera*
   68b Leaves deciduous, pinnately veined. Inflorescence a cyme
     opposite a leaf . . . . *Solanum*
67b Trees or non-climbing shrubs.
   69a Leaves very reduced, sparse or may be absent, the mass of
     visible green material being branches and twigs; flowers pealike,
     yellow.
      70a Plant a mass of intricately and perpendicularly branching
        stems and green spines (modified twigs) in the axils of minute,
        spine-tipped leaves . . . . *Ulex*
      70b Plant more sparingly and obliquely branching, with green,
        spineless twigs.
        71a Twigs slender, angled in cross section, ascending nearly
          parallel; keel of flower obtuse or rounded . . . . *Cytisus*
        71b Twigs stouter, cylindrical in cross section, ascending
          obliquely; keel of flower acutely pointed . . . . *Spartium*
   69b Leaves of normal size and abundance, present throughout the
     growing season; twigs variously coloured, generally not green.
      72a Stipules united into a sheath around the stem . . *Polygonum*
      72b Stipules, if present, are separate.
        73a Bud enclosed in a single bud scale . . . . *Salix*
        73b Bud with more than one bud scale, or none.
          74a Tree, with lowest bud scale directly above petiole
          base or leaf scar. Pith of twig 5-angled in cross section
          . . . . *Populus*
          74b Lowest scale in one side of bud, or pith not 5-angled,
          or a shrub.
            75a Leaves evergreen, usually tough and leathery.
              76a Leaves entire.
                77a Tree with smooth, orange to reddish, papery
                peeling bark . . . . *Arbutus*
                77b Shrubs.
                  78a Stems trailing on ground.
                    79a Leaves more than 1 cm long, obovate
                    to oblanceolate . . . . *Arctostaphylos*
                    79b Leaves 1 cm or less long, ovate, acute-
                    tipped.
                      80a Leaves with minute dark bristles
                      beneath. Flowers urn-shaped. Fruit a
                      whitish berry . . . . *Gaultheria*

80b Leaves glabrous. Flowers with 4 recurved petals. Fruit a
red to purplish berry . . . . *Vaccinium*
78b Stems ascending.
81a Leaves oblanceolate, widest near tip, dark green, the tip
tending to turn down . . . . *Daphne*
81b Leaves elliptic to lanceolate, widest about mid-length or
below.
82a Leaves and twigs scurfy, with tiny scales.
83a Branchlets elongate, slender, somewhat arching.
Shrub of fens . . . . *Chamaedaphne*
83b Branchlets relatively short and stout. Small upland
shrub . . . . *Rhododendron*
82b Leaves and twigs not scurfy.
84a Leaf margins rolled under.
85a Twigs, buds and undersides of leaves covered with
pale to brown curly hairs . . . . *Ledum*
85b Twigs and undersides of leaves glabrous or with
few minute bristles.
86a Leaves lanceolate, greyish green above, white
beneath. Young buds white . . . . *Andromeda*
86b Leaves elliptic; bright, shiny green above, pale
green beneath . . . . *Vaccinium vitis-idaea*
84b Leaf margins flat or nearly so.
87a Leaves ovate, greyish green equally on both sides.
Young twigs pale green with stiff hairs. Older branches dark reddish brown, profusely branched and
crooked . . . . *Arctostaphylos columbiana*
87b Leaves lanceolate to narrowly elliptic, distinctly
paler beneath than above. Young twigs brownish,
glabrous.
88a Leaves 8-20 cm long, glabrous when mature,
non-glandular. Big, coarse shrub with large pink
flowers . . . . *Rhododendron macrophyllum*
88b Leaves 3-8 cm long, very shortly hairy and
glandular beneath. Modest-sized shrub with smaller
white flowers . . . . *Ledum glandulosum*
76b Leaves toothed or lobed.
89a Leaf teeth spine-tipped . . . . *Ilex*
89b Leaf teeth not spine-tipped.
90a Leaves grey-tomentose and soft; wedge-shaped and apically
truncate, toothed or lobed . . . . *Artemisia*

90b　Leaves leathery and tough, deep green, not apically truncate or lobed.

 91a　Leaves oblanceolate, with few teeth toward apex.

  92a　Tall shrub of the seashore, with dark branches. Leaves with dark spots . . . . *Myrica californica*

  92b　Very low plant with green twigs. Leaves without dark spots . . . . *Chimaphila*

 91b　Leaves ovate to elliptic.

  93a　Leaves with 3 strong veins from base . . . . . . *Ceanothus*

  93b　Leaves pinnately veined.

   94a　Leaves elliptic, 5-12 cm long; lateral veins 10-20 each side . . . . *Arbutus*

   94b　Leaves ovate to lanceolate; lateral veins 2-7 each side.

    95a　Leaves at least twice as long as wide, 2-4 cm long; veins obscure . . . . *Vaccinium ovatum*

    95b　Leaves mostly less than twice as long as wide, of diverse sizes; veins distinct . . . . *Gaultheria*

75b　Leaves deciduous, usually soft and membranous.

 96a　Leaves entire.

  97a　Leaves, twigs and buds scurfy, with minute scales . . . . *Elaeagnus*

  97b　Leaves and twigs not scurfy.

   98a　Twigs with chambered pith . . . . *Oemleria*

   98b　Twigs with solid or spongy pith.

    99a　Twigs creamy white. Leaf midvein fading out toward leaf apex . . . . *Lycium*

    99b　Twigs green to brown. Midvein often ending in a minute projecting knob

     100a　Twigs and underside of leaves with brown or yellowish hairs.

      101a　Leaves narrowly elliptic, 5-9 cm long. Flowers white from separate buds below leaves . . . . *Rhododendron albiflorum*

      101b　Leaves ovate, acuminate, 1-4 cm long. Flowers pink on ends of short lateral, leafy branchlets . . . . *Cotoneaster*

     100b　Twigs and leaves with white hairs or none.

      102a　Leaves oblanceolate, 3-6 cm long, with 2 or 3 narrowly oblique lateral veins either side, glabrous and shiny. Flower solitary, terminal, wide open, copper-red . . . . *Cladothamnus*

102b   Leaves ovate to oblanceolate; if oblanceolate, smaller
than above and not shiny. Flowers small, axillary, urn-shaped.
103a Calyx and corolla lobes 5 each, ovary inferior, form-
ing a berry . . . . *Vaccinium*
103b Calyx and corolla lobes 4 each, ovary superior,
forming a capsule . . . . *Menziesia*
96b   Leaves toothed or lobed.
104a   Leaves palmately lobed and veined.
105a   Stems prickly.
106a   Leaves more than 15 cm wide, prickly . . . . *Oplopanax*
106b   Leaves less than 15 cm wide, unarmed . . . . *Ribes*
105b   Stems unarmed.
107a   Stems with noticeably shredding bark. Leaves usually
3-lobed . . . . *Physocarpus*
107b   Stems with relatively firm bark.
108a   Leaves with lanceolate stipules. Flower clusters terminal . . . . *Rubus parviflorus*
108b   Leaves without stipules or at most a few bristles.
Flowers in axillary racemes . . . . *Ribes*
104b   Leaves normally pinnately veined; pinnately lobed or
unlobed.
109a   Stems prickly or thorny.
110a   Plant with small hooked prickles scattered along stem
and leafstalks . . . . *Rubus nivalis*
110b   Plant with straight spines confined to nodes on stem.
111a   Spine in place of leaf, with a leaf tuft in its axil
. . . . *Berberis*
111b   Spine or spur axillary or above a leaf scar.
112a   Spine smooth, sharp-pointed, without leaves or
buds . . . . , . *Crataegus*
112b   Spine or spur with leaves and buds, at least when
young; later with leaf and bud scars.
113a   Leaves unlobed, oblanceolate or obovate, narrowly tapering to base, 2-4 cm long. Fruit a small
black plum . . . . *Prunus spinosa*
113b   Leaves lobed.
114a   Leaves deeply lobed; spurs needle-sharp,
with lateral buds only . . . *Crataegus monogyna*
114b   Some leaves with 1 or 2 shallow lobes near
base; spurs stiff, blunt, with terminal buds
. . . . *Malus diversifolia*

109b Stem not prickly, although may have widely projecting spurs that are not spiny-tipped.

115a Leaves wedge-shaped, 1-3 cm long, with usually 3 teeth across the truncate apex .... *Purshia*

115b Leaves otherwise.

116a Prostrate, mat-forming shrubs.

117a Leaves obovate, tapering to base, glabrous, with flat, crenate margin .... *Arctostaphylos alpina*

117b Leaves lanceolate to ovate, rounded to subcordate at base, white-hairy beneath, the margin rolled under ... *Dryas*

116b Tall, erect or ascending shrubs or trees.

118a All or some leaves lobed.

119a All leaves prominently pinnately lobed, but not serrate. Buds clustered toward end of twig. Pith 5-angled in cross-section .... *Quercus*

119b Leaves serrate as well as lobed.

120a Sinuses between lobes rounded, untoothed; some leaves unlobed .... *Morus*

120b Sinuses V-shaped.

121a Leaves with one lobe near base, or none in some leaves .... *Malus*

121b Leaves with 3 or more toothed lobes on each side .... *Holodiscus*

118b Leaves with teeth but not lobed.

122a Leaves strongly 3-veined from base .... *Ceanothus*

122b Leaves normally pinnately veined.

123a Glands at top of leaf stalk. Buds tending to be clustered at end of twig. Fruit a stone fruit ... *Prunus*

123b Leaf stalks glandless. Buds usually not clustered.

124a Leaves oblanceolate, tapering to base, with few small teeth near tip, dotted with tiny yellow glands. Aromatic shrub of shores and swamps .... *Myrica*

124b Leaves ovate to obovate, serrate full length or with conspicuous teeth toward apex. Glands absent except in *Betula*.

125a Leaves distinctly asymmetric, 2-ranked, sharply serrate from base, the many lateral veins ending in teeth. Branchlets light grey or brownish. Fruit a green circular samara on a stalk .... *Ulmus*

125b Leaves quite symmetric. Branchlets darker. Fruit otherwise.
126a Stipules present, at least early in growth, attached to twig and leaving a scar after falling.
127a Main lateral veins curving to nearly parallel to leaf margin, not running straight into teeth . . . . *Rhamnus alnifolia*
127b Main lateral veins running straight into teeth.
128a Bud enclosed by 2 scales. Terminal bud present
. . . . *Alnus*
128b Bud scales more than 2. Terminal bud absent.
129a Twigs yellowish brown. Leaves obovate and doubly serrate, acuminate at tip. Fruit a nut in an enclosing involucre . . . . *Corylus*
129b Twigs reddish to greyish brown. Leaves ovate and acuminate, or if obovate or circular, neither doubly serrate nor acuminate at apex . . . . *Betula*
126b Stipules absent, or if present attached to leaf stalk, leaving no scar on twig after falling.
130a Branches bearing well-developed, stout lateral shoots that bear leaves and end in inflorescences.
131a Leaves usually less than 4 cm long, broadly elliptic, obtuse or rounded at apex, few-toothed. Flowers in racemes, white with narrow petals. Fruit juicy, blue-black, less than 1 cm in diameter . . . . *Amelanchier*
131b Leaves usually more than 5 cm long, narrowly elliptic, acute at apex, many-toothed.
132a Twigs hairy; bark brown, breaking into long strips or irregular plates; fruit an apple . . . . *Malus*
132b Twigs mostly glabrous; bark dark grey to blackish, breaking into small rectangular plates; fruit a pear
. . . . *Pyrus*
130b Branches not bearing stout lateral shoots. Inflorescences terminal or axillary.
133a Leaves rather coarsely few-toothed, mainly near apex. Flowers many, very small, in terminal panicles . . . . *Spiraea*
133b Leaves finely many-toothed along full length. Flowers with urn-shaped corolla.
134a Terminal bud present. Branches arising in a cluster from the end of each year's stem growth. Flowers in a terminal cluster, with 4-lobed corolla . . . . *Menziesia*
134b Terminal bud absent. Branches from scattered buds along previous year's stem growth. Flowers 1 or 2 from a leaf axil, with 5-lobed corolla . . . . *Vaccinium*

TREES AND SHRUBS

Pinaceae   Pine Family

Typically tall, columnar trees, with massive, erect central trunks and short, slender lateral branches. Leaves alternate or in alternately arranged bundles; linear, needlelike, evergreen except in *Larix,* with a single central vascular (conducting) strand or vein (doubled in the "hard" pines). Pollen cones with stamens with 2 pollen sacs. Seed cones bearing spirally arranged bracts (sometimes very small) basally attached to the ovule- and seed-bearing cone scales in their axils. Each scale bears 2 inverted ovules. In most species the seeds are winged and are shed as the cone scales spread apart to release them. Seedlings with 3 or more cotyledons. Plates 1-8.

Key to the Genera of Pinaceae

1a Leaves attached in bundles.
   2a Leaves evergreen, in bundles of 2, 3 or 5 . . . . *Pinus*
   2b Leaves deciduous, in bundles of more than 10 . . . . *Larix*
1b Leaves attached singly.
      3a Leaves more or less 4-angled, attached to the twig on projecting woody bases that remain on the twig after leaf fall . . . . *Picea*
      3b Leaves more or less flat. Twigs without prominent persisting bases.
         4a Mature cones pendant, with persistent cone scales. Leaf scars oval.
            5a Bracts projecting from cone, as long as the scales, 3-pronged. Twigs smooth. Terminal bud conspicuous, acute . . . . *Pseudotsuga*
            5b Bracts shorter, not projecting from mature cones. Twigs fluted. Terminal bud very small, blunt . . . . *Tsuga*
         4b Mature cones erect, with deciduous scales. Leaf scars circular . . . . *Abies*

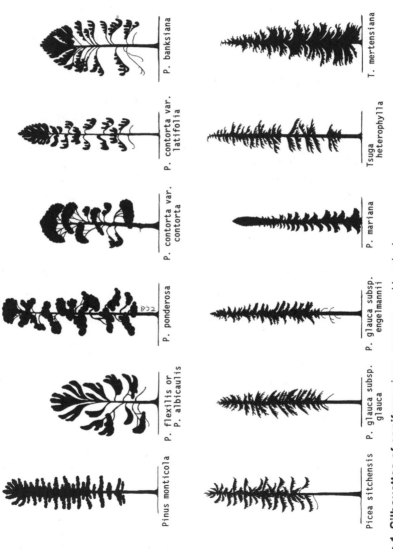

Plate 1. Silhouettes of conifers: pines, spruces and hemlocks.

P. banksiana

P. contorta var. latifolia

P. contorta var. contorta

P. ponderosa

P. flexilis or P. albicaulis

Pinus monticola

T. mertensiana

Tsuga heterophylla

P. mariana

P. glauca subsp. engelmannii

P. glauca subsp. glauca

Picea sitchensis

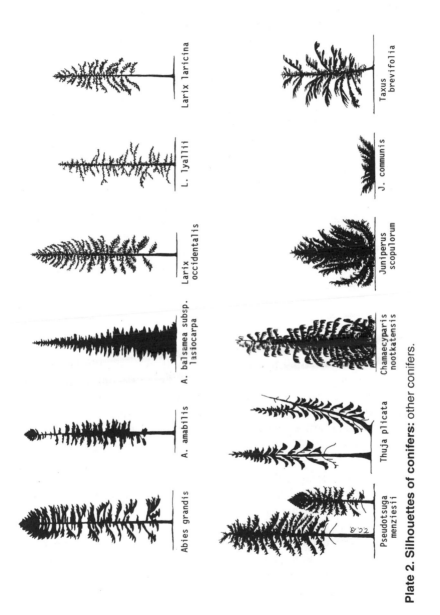

Plate 2. Silhouettes of conifers: other conifers.

Larix laricina

L. lyallii

Larix occidentalis

A. balsamea subsp. lasiocarpa

A. amabilis

Abies grandis

Taxus brevifolia

J. communis

Juniperus scopulorum

Chamaecyparis nootkatensis

Thuja plicata

Pseudotsuga menziesii

Picea   Spruces

Tall trees with thin, scaly bark. Leaves spirally arranged, extending from all sides of the twigs, persisting for 7-10 years, entire, jointed to a prominent, persistent woody base (sterigma). Cones maturing the first autumn, often terminal; pendent except in Black Spruce; scales thin, much longer than the bracts, persistent on the axis of the cone. Seeds much shorter than the wings. Seedlings with 4-15 cotyledons.

The spruces, except *P. mariana,* are commercially important in British Columbia.

Key to the Species of *Picea*

1a Cones persisting several years, usually less than 3 cm long, purplish
. . . . *P. mariana*
1b Cones deciduous, usually more than 3 cm long, brown when mature.
 2a Leaves more or less flat . . . . *P. sitchensis*
 2b Leaves distinctly 4-sided.
  3a Cone scales almost circular; rounded and entire at apex; stiff.
  Twigs yellowish, glabrous . . . . *P. glauca* subsp. *glauca*
  3b Cone scales rhombic; erose at apex; papery and flexible. Twigs
  brownish, minutely puberulent . . . . *P. glauca* subsp. *engelmannii*

Picea sitchensis (Bongard) Carriere   Sitka Spruce

Largest of all spruces, commonly 40-55 m tall and 1-2 m in trunk diameter (exceptionally to 80 m tall and 3 m in diameter), with dark brown, coarsely scaly bark; twigs glabrous, pale green becoming yellowish. **Leaves** sharp-pointed, stiff, flat in cross section, 15-25 mm long. **Cones** cylindrical, 5-8 cm long, falling soon after maturity; scales pale yellowish, papery, oblong, erose; the lanceolate bracts half as long as the scales. Plate 3, figure 1.

Range: From Alaska southward along the coast and west of the Coast and Cascade ranges to northern California, from the beaches and alluvial flats to passes in the mountains; spilling over into the Nass Basin, where it mingles with *P. glauca* subsp. *engelmannii* (22).

More tolerant than our other conifers of the impact of salty spray blown off the sea, this species commonly forms pure stands in a narrow belt facing the ocean along the exposed outer coast of Vancouver Island and other islands.

Picea × *lutzii* Little (Roche Spruce) is a hybrid (*Picea glauca* × *sitchensis*) found in Alaska and northwestern British Columbia, where it has been found at Kelsall Lake, Hazelton and Stewart.

Picea glauca (Moench) Voss subsp. *glauca*   White Spruce

Tree up to 30 m tall, with a steeply conical to columnar crown and ashy brown, closely scaly bark; twigs glabrous, pale glaucous green becoming yellowish, the persisting leaf bases projecting less than 1 mm from the twig. **Leaves** 4-angled in cross section, 15-18 mm long, rather blunt-tipped, dark greyish green. **Cones** cylindrical, 3-5 cm long, dull reddish brown; scales broadly obovate to circular, ratio of width to length about 0.8; broadly rounded and entire at apex, stiff; bracts small, round to spatulate, a quarter to a third as long as the scale, finely toothed at apex. Seedlings with 6-9 cotyledons.
Plate 3, figure 3.

Key to the Varieties of *P. glauca*
1a Bark smooth, light grey, with obvious resin blisters . . . . **var.** *porsildii*
1b Bark scaly, darker; resin blisters not obvious.
    2a Cones narrowly ovoid, 4-5 cm long; scales longer than broad, flexible, light brown. Crown conical (Photo 1) . . . . **var.** *glauca*
    2b Cones broadly ovoid, 3-3.5 cm long, often nearly as broad as long; scales dark brown, rigid. Crown narrow, columnar (Photo 2)
    . . . . **var.** *albertiana*

Range: Transcontinental in the Boreal Forest Region (53) from the northern limit of tree growth southward into the central interior of British Columbia and along the Rocky Mountains to Montana, mostly at altitudes between 300 and 1,300 m, mostly on well-drained upland soils, and along streams at its lower altitudinal limit. Var. *albertiana* (Stewardson-Brown) Sargent (Alberta White Spruce) is our common variety in British Columbia. Var. *porsildii* Raup (Porsild Spruce) occurs with var. *albertiana* in the Rocky Mountains.

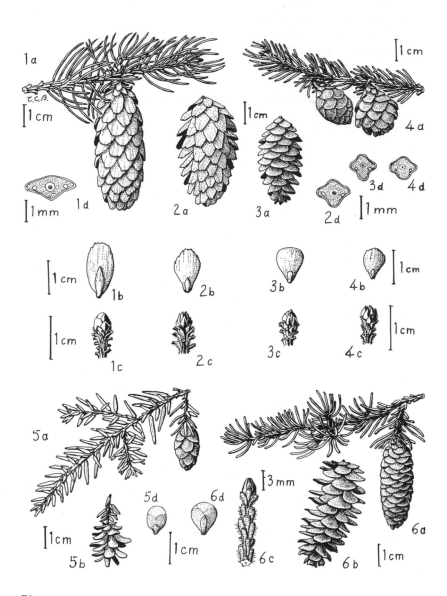

Plate 3 Spruces and Hemlocks. Figures: (1) *Picea sitchensis;* (2) *P. glauca* subsp. *engelmannii;* (3) *P. glauca* subsp. *glauca;* (4) *P. mariana;* (5) *Tsuga heterophylla;* (6) *T. mertensiana.* For *Picea:* b, cone scales and bracts; c, twigs and buds; d, cross sections of leaves. For *Tsuga:* b, open cones; c, twig and bud; d, cone scales and bracts.

Picea glauca subsp. *engelmannii*   Engelmann Spruce
(Parry *ex* Engelmann) Taylor
(*P. engelmannii* Parry *ex* Engelmann)

Tree up to 45 m tall and 1.2 m in diameter, with a columnar or slenderly
conical crown, or forming a depressed, shrubby krummholz at alpine levels; bark light reddish brown, more coarsely scaly than in subsp. *glauca,*
but appearing grey and smooth at a distance. Twigs slender, brownish,
finely hairy with brownish hairs; persistent leaf bases often projecting a
millimetre or more from the twig surface. **Leaves** 4-angled in cross section, up to 30 mm long, acute at tip. **Cones** ovoid, up to 6.5 cm long, sessile or very short-stalked, soft, paler and more yellowish brown than
White Spruce cones; scales elliptic to rhombic, ratio of width to length
about 0.6, thin and flexible, the margin erose and often wavy, the apex
obtuse to narrowly truncate; bracts broadly lanceolate, acute and often
erose at apex, about a third the length of the scale. Seedlings usually with
6 cotyledons. Plate 3, figure 2. Photo 3.

Range: Cordilleran region from central British Columbia and western
Alberta along the mountain ranges to Arizona and New Mexico. In British
Columbia, from the east slope of the Coast and Cascade ranges to the
Rocky Mountains, northward to about lat. 55°N in the Skeena Basin;
between 1,000 m altitude and timberline; overlapping geographically and
altitudinally with subsp. *glauca.* These subspecies of *P. glauca* intergrade
in various forms throughout the central interior of British Columbia and
the Rocky Mountains of Alberta (57).

Picea mariana (Miller) B.S.P.   Black Spruce

Usually a smallish tree with a slender columnar crown of branches that
turn downward and outward, rather coarsely scaly bark, and brownish,
finely hairy twigs. **Leaves** 4-angled in cross section, 6-15 mm long, blunt-
tipped, dark greyish green. **Cones** ovoid, 13-40 mm long, often in persistent clusters near the trunk toward the top of the tree, their bases reflexed;
the scales purple, stiff, rounded, with entire or ragged margins. The cones
may persist on the tree for many years after ripening. Plate 3, figure 4.

Range: Transcontinental in the Boreal Forest Region, often dominant in
cold swamps and bogs (Photo 4), but not confined to such habitats. In

British Columbia, common north of the upper Fraser River, southward to about 15 km north of Quesnel, across northern British Columbia, but nowhere approaching the coast or even the Coast Range in this province.

Tsuga   Hemlocks

Tree, often with nodding or drooping, whiplike leading shoots, and spreading, declining branches. Twigs slender, with low, cushionlike leaf bases and oval leaf scars. Leaves lanceolate to linear, short-stalked, often variable in length on one branch, flat. Cones pendulous, on terminal or short lateral branchlets; the scales spreading widely at maturity, the bracts shorter than the scales. Seedlings with 3-6 cotyledons.

Key to the Species of *Tsuga*
1a Leaves very diverse in length, flat, rounded at apex; with stomata only on the lower surface. Cones ovoid, not more than 25 mm long
. . . . *T. heterophylla*
1b Leaves nearly uniform in length, plump, bluntly pointed at apex, with stomata on both surfaces. Cones cylindrical, more than 25 mm long
. . . . *T. mertensiana*

Tsuga heterophylla (Rafinesque) Sargent   **Western Hemlock**
(*T. mertensiana* [Gordon] Carriere)

Tall tree of narrow columnar form with slender branches and declining or drooping branchlets, up to 60 m tall and 2 m in diameter. Bark russet-brown, scaly, thinner than in Douglas-fir, ultimately becoming shallowly furrowed into flat ridges; young twigs slender, pubescent. **Leaves** appearing 2-ranked, of varying lengths on a twig; leaves arising from the underside of the twig are notably longer than those arising above them; short-stalked; glaucous beneath in two bands in which the stomata are visible as minute dots; upper surface green, without stomata. **Cones** ovoid, 20-25 mm long, maturing in one season, sessile and pendent at ends of short lateral shoots; opening wide at maturity; scales longer than wide, broadly rounded and entire at apex; bracts small, broadly triangular, rounded at apex. Plate 3, figure 5.

Range: Southern Alaska to California along the coast, including Vancouver Island and the Queen Charlotte Islands, and inland in the Columbia Forest Region, from the Parsnip River at lat. 56°N through the humid mountain valleys to Idaho and Montana. It becomes a climax co-dominant with *Abies amabilis* on the humid western slopes of Vancouver Island.

Tsuga mertensiana (Bongard) Carriere Mountain Hemlock
(*T. pattoniana* [Murray] Engelmann)

Tree up to 40 m tall and 120 cm in diameter, with thick, dark brown bark fissured into flat, platy ridges, and stiff branches declining and curving outward. **Leaves** spirally attached on long shoots, crowded on short ones, more uniform in length than those of *T. heterophylla,* without a glaucous surface but with stomata on both surfaces, blunt-pointed, often curved. **Cones** 5-7.5 cm long, cylindrical, often in clusters; scales thin, rounded at apex, as wide as long. Plate 3, figure 6. Photo 5.

Range: Southern Alaska to California along the coastal mountain ranges, and inland from the Monashee, Selkirk and Purcell ranges southward into Montana and Idaho. In British Columbia, mainly in humid Subalpine Forest zones; replacing *T. heterophylla* above 1,000 m on Vancouver Island and above 1,400 m in the Selkirk Mountains.

Pseudotsuga

Evergreen trees with thick, dark brown, fissured bark. Leaves flat, alternate and spirally attached, but sometimes appearing 2-ranked by a twist of the leaf bases; petiols short, with 2 resin ducts, and when shed, leave an oval scar flush with the twig surface. Cones monoecious, axillary, with prominent 3-pronged bracts longer than their scales; maturing in one season; erect when young, pendent when ripe. Seeds nearly straight on one side, obliquely rounded on the other, with a wing attached to the convex side. Seedlings with 6-12 cotyledons.

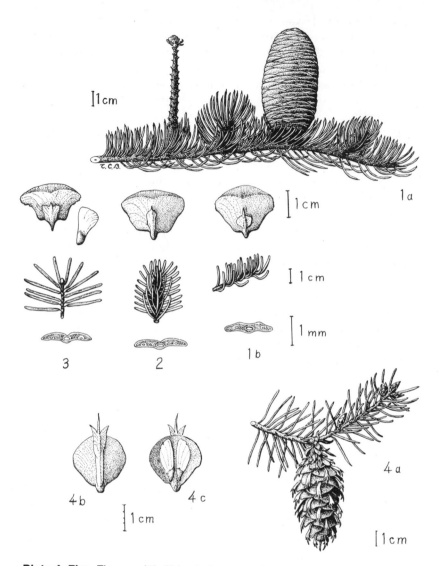

Plate 4 Firs. Figures: (1) *Abies balsamea* subsp. *lasiocarpa:* 1a, upper branch with cone and old cone axis; 1b, cone scale (dorsal view) with bract, lower branch foliage, and leaf cross section, showing position of resin ducts; (2) *A. amabilis:* cone scale, foliage, and leaf section; (3) *A. grandis:* cone scale, seed (to same scale), foliage, and leaf section; (4) *Pseudotsuga menziesii:* 4b, cone scale with bract, dorsal view; 4c, ventral view of same, with attached seeds.

Pseudotsuga menziesii (Mirbel) Franco   **Douglas-fir**
(*P. taxifolia* [Lambert] Britton)

Large tree, up to 60 m tall (exceptionally over 75 m) with a trunk diameter up to 1.8 m (exceptionally up to 4 m); the bark dark brown, thick, deeply fissured longitudinally. **Leaves** 20-30(-40) mm long, short-stalked, linear, bluntly pointed. **Cones** ovoid, 4-10(-15) cm long; scales broadly elliptic to circular, the bracts longer than the scales. Seeds 6 mm long with a wing twice as long. Seedlings with around 7 cotyledons. Plate 4, figure 4. Photo 6.

Range: British Columbia and southwestern Alberta to Mexico through the Cordilleran region. In British Columbia, along the coast northward to lat. 53°30'N on the Kemano River; on the Skeena River at latitude 55°N according to G.M. Dawson (9), but not seen there in recent years. In the interior, northward to lat. 55°30'N at Babine, Takla and Tudyah lakes; eastward to the Rocky Mountains, from sea level up to around 1,500 m altitude in the southern interior. A climax dominant of the Coast Forest at low elevations around the Strait of Georgia and of the Montane Forest on the Interior Plateau. Var. *menziesii,* with bracts slightly longer than the scales, occurs along the coast. Var. *glauca* (Beissner) Franco, distinguished by relatively longer, diverging bracts, is the common variety in the interior.

The largest living Douglas-fir currently on record is the Red Creek tree near Port Renfrew (51). It is 4 m in diameter and 74 m tall to broken top. Ages of over 1,300 years have been reported.

Abies   **Firs**

Trees with columnar or spirelike crowns, resinous bark and stiff horizontal branches mainly in whorls, bearing 2-ranked branchlets, forming flat masses of foliage. Twigs with circular leaf scars flush with the twig surface. Buds in ours are encased in resin. Leaves linear, with very short, stout stalks. Cones are monoecious. Staminate cones numerous in leaf axils beneath the twigs. Ovulate cones stand erect on the upper branches, mature in one season, then disintegrate, leaving the central axis persisting; scales thin, very broadly cuneate, with a rounded entire apex and a stalked base; bracts small in ours. Seedlings with 4-7 or more cotyledons.

Key to the Species of *Abies*

1a Leaves, at least on the lower branches, spreading laterally from twigs; commonly notched at apex; dark green above, white beneath, with stomata; resin ducts against the lower leaf surface.

    2a All leaves spreading in one plane; cones green; bracts with humped shoulders and short, pointed tip . . . . *A. grandis*

    2b Some leaves lying along top of twig; cones purple; bracts oblanceolate, tapered to the tip . . . . *A. amabilis*

1b Leaves all turned upward, the apex acute or rounded, greyish green and bearing stomata on both surfaces; resin ducts medial, not against lower leaf surface . . . . *A. balsamea* **subsp.** *lasiocarpa*

Abies amabilis (Douglas *ex* Loudon) Forbes   Amabilis Fir

Tree with a slender columnar crown, up to 50 m tall and 120 cm in diameter. Bark very resinous, smooth with white markings, becoming fissured into small plates on old trees. **Leaves** on lower branches spreading, with those on the upper side of the twig lying along the twig, 20-30 mm long, the upper surface grooved, the tip usually notched; leaves on upper branches shorter and sharper, tending to curve upward. **Cones** purple, becoming brown when ripe, barrel-shaped, 10-16 cm long by up to 6 cm thick; scales about 25 mm wide and nearly as long; bracts oblanceolate, tapering to base and to an acute tip. Seeds light brown, wrinkled, 9 mm long. Plate 4, figure 2. Photo 8.

Range: From southeastern Alaska to California along the coast and the western slopes of the Coast and Cascade ranges. In British Columbia, inland to Manning Provincial Park; common on Vancouver Island, from sea level on the outer coast to 1,500 m altitude, on the north coast up to 300 metres; absent from the Queen Charlotte Islands. A tree of the most humid rain forests (50), this species is a climax co-dominant with Western Hemlock on well-drained slopes on western Vancouver Island at low elevations and with Mountain Hemlock at subalpine levels.

Abies grandis   **Grand Fir**
(Douglas *ex* D. Don in Lambert) Lindley

Tree of columnar form, 35-45(-70) m tall and 60-120(-150) cm in diameter. All but the topmost branches spread stiffly horizontally, to slightly downward near the base. Bark thin and smooth, changing earlier than in *A. amabilis* to narrowly furrowed between long, flat ridges. **Leaves** on lower branches 35-50 mm long; grooved and dark green above, notched at apex; spreading from the twig in a horizontal plane. On upper branches, leaves are shorter, thicker and crowded, and turn upward. **Cones** light green, cylindrical, 8-13 cm long and up to 4 cm thick; bracts high-shouldered, with a short, pointed apex. Seedlings usually with 6 cotyledons. Plate 4, figure 3. Photo 7.

Range: Southern British Columbia to California along the coast, inland to the Cascade Range, and from about lat. 50°N in the Kootenay Region southward into Idaho, Montana and eastern Oregon. In British Columbia, common at low elevations on Vancouver Island and the mainland coast south of Johnstone Strait at about lat. 50°30'N, and inland to the upper Skagit Valley at 900 m elevation; and also in the South Kootenay region.

Abies balsamea **(L.) Miller**   **Alpine Fir**
subsp. *lasiocarpa* (Hooker) Boivin   **Subalpine Fir**
(*A. lasiocarpa* [Hooker] Nuttall)

Tree with a distinctly narrow, spirelike form, up to 30 m tall and up to 70 cm in trunk diameter; or a depressed krummholz form at alpine levels; bark thin, smooth and grey, or occasionally thick and furrowed. **Leaves** 20-30(-40) mm long, rounded to blunt at tips on the lower branches, but sharp-pointed on upper, cone-bearing branches; curving upward, rarely spreading; greyish green and stoma-bearing on both surfaces, with resin ducts on either side of the midvein and midway between the upper and lower surfaces. **Cones** 5-10 cm long by 2.5-4 cm thick, erect, deep greyish purple; scales tapering from broad rounded apices; bracts round, with a slender projecting tip. Seedlings usually with 4 cotyledons around 13 mm long. Plate 4, figure 1. Photo 9.

Range: From Yukon and a few, scattered localities in Alaska and the Northwest Territories, through British Columbia and western Alberta, southward via the Cascade Range to Oregon and via the Rocky Mountains to Arizona and New Mexico. Widespread across British Columbia, including Vancouver Island but not the Queen Charlotte Islands. Characteristic of the Subalpine Forest, above 600 metres in the north and 1,200 m in the south; often forming a dense shrubby krummholz at timberline (Photo 11).

Subsp. *balsamea*, which ranges from central Alberta to Newfoundland, differs from subsp. *lasiocarpa* mainly in the more spreading foliage and in the absence of stomata on the upper leaf surface, which is usually a darker green (2).

Larix Larches

Conical, deciduous trees, with slender branches; thick, scaly or furrowed bark; and heavy wood. Branchlets slender, often somewhat pendulous, bearing short stout, spurlike lateral branchlets. Buds small, rounded, covered with broad scales. Leaves alternate and spaced out along young terminal twigs, but densely tufted on the lateral spurs, all falling each year. Fruit a cone with thin scales and wide-based, slender-tipped bracts, maturing in one year. Seeds nearly triangular, rounded on the sides, light brown, shorter than their wings. Seedlings with 6 cotyledons.

Key to the Species of *Larix*
1a Cones usually less than 25 mm long; bracts shorter than the scales;
    leaves 25 mm long .... *L. laricina*
1b Cones usually longer than 25 mm; bracts longer than the scales;
    leaves 30-40 mm long.
      2a Twigs densely hairy; leaves 4-angled in section .... *L. lyallii*
      2b Twigs sparsely hairy or glabrous; leaves triangular in section
        .... *L. occidentalis*

Larix laricina (DuRoi) K. Koch   Tamarack
(*L. americana* Michaux)

Tree up to 20 m tall, with reddish to greyish brown, scaly bark; twigs glabrous, glaucous. **Leaves** in clusters of 12-50, approximately 25 mm long, with 2 minute resin ducts, light green. **Cones** usually less than 25 mm long, obovoid, the bracts shorter than the few elliptic scales. Plate 5, figure 3.

Range: Central Alaska to the Atlantic coast in the Boreal Forest Region. In British Columbia, mainly east of the Rocky Mountains and through the Liard River Basin, with an isolated area west of Prince George and southward to Pantage Lake, at lat. 53°10'N, long. 123°05'W. Typically found in and around swamps and Sphagnum bogs.

Larix lyallii Parlatore in DC.   Alpine Larch

Small to medium-sized tree, often of irregular form; with reddish brown, appressed-scaly bark; the twigs and buds finely and densely puberulent. **Leaves** in bundles of 30-50, light bluish-green, 3-4 cm long, 4-angled in cross section, with 2 resin ducts. **Cones** cylindrical, 4-5 cm long; the scales obovate, hairy on the outer sides; bracts ovate-based, with long, tapering, exserted tips that recurve on the ripe cones. Plate 5, figure 1. Photo 10.

Range: Southern British Columbia, southwestern Alberta, Washington, Idaho and western Montana. In British Columbia, on rocky sites, often on north slopes, in subalpine and timberline areas of the Rocky Mountains northward to around the Kicking Horse Pass in the southern Selkirk range, and in the Cascade and Okanagan ranges south of the Similkameen River.

Larix occidentalis Nuttall   Western Larch

A large tree, 30-45 m or more in height, with thick, reddish brown, scaly bark; twigs sparsely hairy at first, to glabrous. **Leaves** yellow-green, 4-5 cm long, in bundles of 12-30 or more, triangular in cross section, without resin ducts. **Cones** ovate, 2.5-4 cm long, at first purplish-red, later reddish-brown, short-stalked; scales orbicular, ciliate when young; bracts ovate with long, linear, exserted, ultimately recurving tips. Plate 5, figure 2.

Range: East of the Cascade Range, from southern British Columbia and southwestern Alberta to Oregon and Montana. In British Columbia, in moist valleys and north slopes of the Kootenay Region, northward to Revelstoke and Shuswap Lake, westward to the Okanagan Valley and to Guichon Creek (lat. 50°10'N, long. 120°50'W); formerly reported from the Barrière River (lat. 51°15'N, long. 120°W).

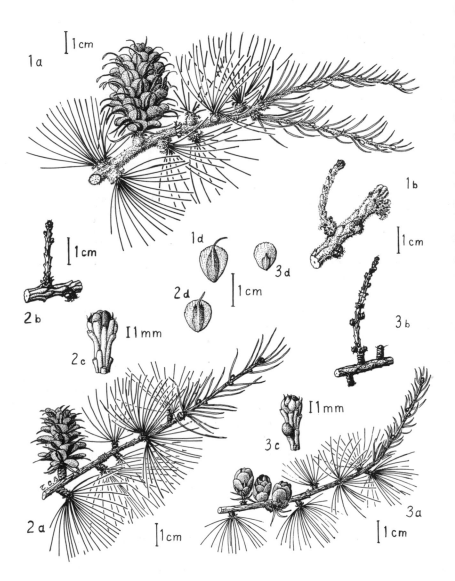

Plate 5 Larches. Figures: (1) *Larix lyallii;* (2) *L. occidentalis;* (3) *L. laricina.*
1b, 2b, 3b, leafless twigs; 2c, 3c, terminal buds; 1d, 2d, 3d, cone scales
and bracts.

Pinus   Pines

Trees with needlelike leaves in bundles of 2-5 in the axils of scale-leaves; the bundles being enclosed in chaffy, basal sheaths at least when young, the sheaths being shed early in the 5-needled species. Cones often conspicuous, becoming woody with terminal thickening of the scales, maturing in their second or third year. Seeds usually winged. Seedlings with 3-18 cotyledons. Plates 1, 6, 7 and 8.

Key to the Species of *Pinus*
1a Leaves in bundles of 2 or 3, their basal sheaths persistent. Bark becoming scaly when still young. The "Hard Pines."
  2a Leaves normally in pairs, 2-8 cm long, persistent.
    3a Cones upcurved, without prickles; east of Rocky Mountains
    . . . . *P. banksiana*
    3b Cones straight or reflexed, prickly; widespread . . . . *P. contorta*
  2b Leaves normally in threes, 12-18 cm long; cones 5-14 cm long, shed the year after ripening. Dry sites in southern interior
    . . . . *P. ponderosa*
1b Leaves in bundles of 5, their basal sheaths shed early. Bark grey, remaining smooth many years, becoming furrowed or platy in age. The "Soft Pines."
  4a Cones cylindrical, 13-30 cm long, stalked, becoming pendulous. Leaf margins serrulate. Moist regions of southern British Columbia . . . . *P. monticola*
  4b Cones ovoid, 4-16 cm long, not pendulous. Leaf margins usually entire.
    5a Cones 8-16 cm long, becoming yellowish and opening when ripe. Rocky Mountains only, in British Columbia . . . . . . *P. flexilis*
    5b Cones 4-8 cm long, dark purple to dark brown, remaining closed when ripe. High altitudes across southern British Columbia
    . . . . *P. albicaulis*

Pinus banksiana Lambert   Jack Pine

Tree 12-15 m tall and 20-30 cm in trunk diameter, with a broadly conical, open crown; long, obliquely ascending and spreading branches; and thick, coarsely flaky bark. **Leaves** in pairs, each pair with a persistent scaly basal sheath, 2-4 cm long, curved and diverging, yellowish-green. **Cones** commonly in threes, asymmetrical and curved up toward the twig tip, normally without prickles, often remaining closed when ripe. Plate 6, figure 4. Photo 14.

Range: From the Mackenzie Valley, Northwest Territories and extreme northeastern British Columbia, eastward to the Atlantic coast in the Boreal Forest Region. In British Columbia, found on the Petitot River, at lat. 59°50'N, long. 122°W, on gravelly and sandy soils.

Pinus contorta Douglas *ex* Loudon — Lodgepole Pine / Shore Pine

Small to medium-sized tree of diverse form. **Leaves** in pairs, occasionally in threes in northwest, with a persistent sheath, 3-8 cm long, semicircular in cross section, the margin serrulate, rather dark yellowish-green. **Cones** straight, standing obliquely outward from the twig, 3-6 cm long, the tips of the scales armed with prickles. Plate 6, figures 2 and 3.

Range: Yukon, Northwest Territories, Alaska and Alberta, southward to Baja California and southeastward to Colorado and South Dakota. Tolerant of a wide range of climatic and soil conditions, from gravel and rocks to saturated swamp soils.

Key to the Varieties of *P. contorta*
1a Crown broad and rounded, on a stout, rather crooked trunk; bark thick, furrowed; cones short, opening when ripe. From coast to Coast range . . . . **var.** *contorta*
1b Crown columnar and slender, on a slender, straight trunk; bark thin, scaly. Cones remaining closed when ripe. All of British Columbia east of Coast range (Yukon to Colorado) . . . . **var.** *latifolia*

Note: Var. *contorta* (Shore Pine) is found in extreme situations with respect to physical environmental factors: in rocky or gravelly, excessively well-drained soils, and in the saturated muck and peat of coastal swamps and bogs. In the latter situations, it often forms open stands of short, stunted treelets. This species' intolerance of competition from larger, more luxuriantly growing species tends to keep it out of the more fertile, intermediate situations, unless it can seed in following a fire, when it forms a pioneer stage in forest succession.

Var. *latifolia* Engelmann *ex* S. Watson (Lodgepole Pine) occurs in almost all forest types in its range and tends to form extensive pure stands of uniform age. This is a conspicuously fire-adapted variety. Its cones remain closed on the tree, holding viable seeds for years until the heat of a fire causes the cone scales to spread and release the seeds into the ashes, to regenerate a new, uniformly aged stand on the burnt area. Without

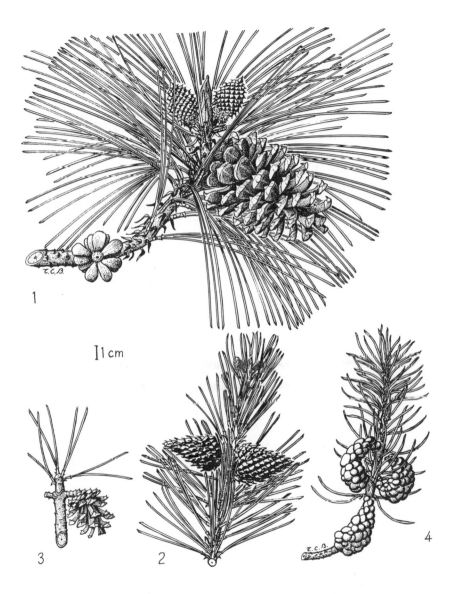

1

I1 cm

3 2 4

Plate 6 The Hard Pines. Figures: (1) *Pinus ponderosa;* (2) *P. contorta* var. *latifolia;* (3) *P. contorta* var. *contorta;* (4) *P. banksiana.*

PINE FAMILY — 63

repeated fires this pine will eventually be replaced by more shade-tolerant species (Photo 13).

Pinus banksiana hybridizes with *P. contorta* var. *latifolia* to produce the hybrid *P.* × *murraybanksiana* Righter & Stockwell. East of the Rocky Mountains, from the Peace River northward and within the range of *P. contorta* var. *latifolia*, but far beyond the present known range of *P. banksiana*, many trees are seen that show characteristics of *P. banksiana*, such as thick bark or long, outwardly arching branches, mixed with characteristics of *P. contorta*, such as straight, prickly cones. This suggests a past episode of interbreeding between these species, at a time when *P. banksiana* must have occupied a much larger area of northeastern British Columbia than it does today.

Pinus ponderosa **Douglas *ex* Lawson**	**Ponderosa Pine**
var. *ponderosa*	**Yellow Pine**
(*P. washoensis* Mason and Stockwell)	

A large tree, variable in form and character, up to 50 m tall and 160 cm in trunk diameter. Bark thick, coarsely scaly and furrowed, dark grey to blackish on young saplings, but becoming pinkish to orange on older trees. **Leaves** 12-26 cm long, usually in bundles of 3 with a scaly basal sheath; leaf cross section a third of a circle, the margin serrulate. **Cones** sessile, ovoid, 5-14 cm long, usually dark purple, rarely green, when almost mature, requiring 2 years to mature; the scales thick, each with a sharp prickle. Seeds with prominent wings. Cones, when shed, leave a ring of scales attached to the twig. Plate 6, figure 1.

Range: British Columbia to Mexico throughout much of the western United States, eastward to Nebraska. In British Columbia, in the dry interior, east of the Cascade and Coast ranges; northward to lat. 51°30'N on the Fraser and North Thompson rivers, westward to the Bridge River Valley, down the Fraser to Ainslie Creek, near Boston Bar, and northward in the Rocky Mountain Trench to Forster Creek, near Radium Hot Springs. It occurs usually between 200 and 1,300 m altitude. At low levels it is a climax dominant, forming open, parklike stands (Photo 12), but at higher elevations it is a pioneer dominant of montane forests dominated ultimately by Douglas-firs (4).

Note: The Washoe Pine (*Pinus washoensis*), recorded from Promontory Hill, near Merritt, at altitudes from 1,400 to 1,700 m, above the usual altitudinal range of Ponderosa Pine in British Columbia, is a high-altitude form of our northern Ponderosa Pine (var. *ponderosa*), with leaves and cones shorter and relatively stouter than the average for Ponderosa Pine. There is no discontinuity in altitude or in the range of variation between these pines. In fact, the range of variation in Washoe Pine falls within that of normal Ponderosa Pine. Leaf length is known to be responsive to external conditions such as the length of the growing season, which is affected by altitude.

Seedling trees grown at sea level from seed collected from the reported Washoe Pines on Promontory Hill produce long, slender, pliable leaves, typical of Ponderosa Pine at low elevations, in contrast to the relatively short, stout, stiff leaves of the parent trees. The leaves require almost 4 months of growth to attain full length.

These observations suggest that the differences between the Washoe and Ponderosa Pines are of environmental origin, rather than genetic; and that the Washoe Pine is really a response of the Ponderosa Pine to the short growing season and to other severe environmental conditions encountered at altitudes above its usual range.

Washoe Pine has been reported in widely scattered stands, on high mountain slopes, from British Columbia to northern California and Nevada. On Promontory Hill it occupies rocky south slopes, above and continuous with stands of normal Ponderosa Pine.

Pinus monticola **Douglas**   **Western White Pine**
ex **D. Don in Lambert**

Tree 30-60 m tall, with a columnar crown. The bark, initially thin, smooth and grey, becomes thick, dark grey and fissured into rectangular plates on mature trunks. **Leaves** in bundles of 5, of which the basal sheaths are deciduous early; the leaves are 5-10 cm long, slender, serrulate, pliable and bluish-green, the cross section a fifth of a circle. **Cones** stalked, more or less cylindrical, 13-30 cm long, without prickles, becoming pendulous, greyish brown when ripe, maturing their second year, then deciduous. Seeds with long wings. Seedlings with 7-10 cotyledons. Plate 7, figure 1.

Range: British Columbia to California and Montana. In British Columbia, along the coast northward to the Homathko River; on Vancouver Island south of Nimpkish Lake; and in the Columbia Forest Region, in the Monashee, Selkirk and Purcell ranges, northward to around Shuswap Lake, at low to moderate elevations.

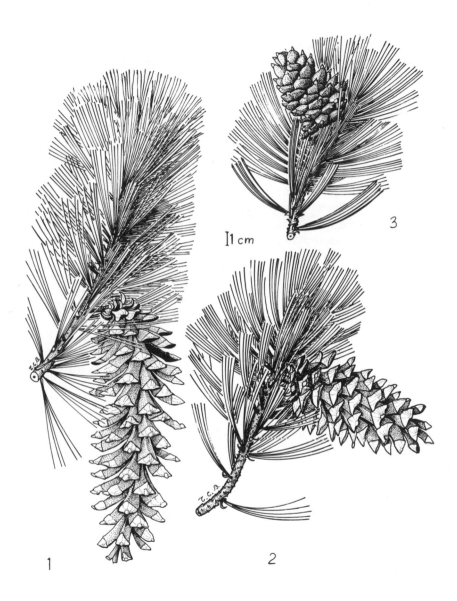

Plate 7 The Soft Pines. Figures: (1) *Pinus monticola;* (2) *P. flexilis;* (3) *P. albicaulis.*

Pinus albicaulis Engelmann   Whitebark Pine

Usually a smallish, rather crooked tree, rarely straight and up to 30 m tall, sometimes forming a dwarf krummholz form at alpine levels; with thin, light grey bark ultimately breaking into oblong plates; the upper crown with long, steeply ascending branches. **Leaves** in sheathless bundles of 5, stout and stiff, 4-8 cm long, entire, greyish green. **Cones** ovoid, 4-8 cm long, dull blackish purple, finally turning brown; closed when ripe; the thick, rather fleshy cone scales with bluntly pyramidal exposed tips. Seeds 10-13 mm long, wingless, embedded in scale bases until removed by birds. Plate 7, figure 3.

Range: From central British Columbia and western Alberta southward along the mountain ranges to California and Colorado. In British Columbia, in the Coast Range northward to about lat. 55°N and in the Rocky and Cariboo mountains northward to around McBride (47); found in open rocky sites in the Subalpine Forest Zone, and at timberline, where often short and stunted.

Pinus flexilis James   Limber Pine

Usually a small tree, similar to Whitebark Pine, with pale grey bark breaking into roundish or square plates on old trees. **Leaves** in sheathless bundles of 5, thick, 4-9 cm long, usually entire, stiff, the leaf cross section a fifth of a circle. **Cones** elongate-ovoid, 6-16 cm long, short-stalked, opening at maturity, yellowish to light brown; lustrous; scales thickened and ridged at apex, with a darker, obtuse tip, the lowest scale tips elongat ed and reflexed. Seed with or without an abbreviated wing. Plate 7, figure 2. Photo 15.

Range: Western Alberta to the Dakotas, the Rocky Mountain states and California. In British Columbia, uncommon on the western slope of the Rocky Mountains: found at Canal Flats, Radium Hot Springs and Golden. Occurring on open rock slopes at moderate elevations (1,000-1,200 m).

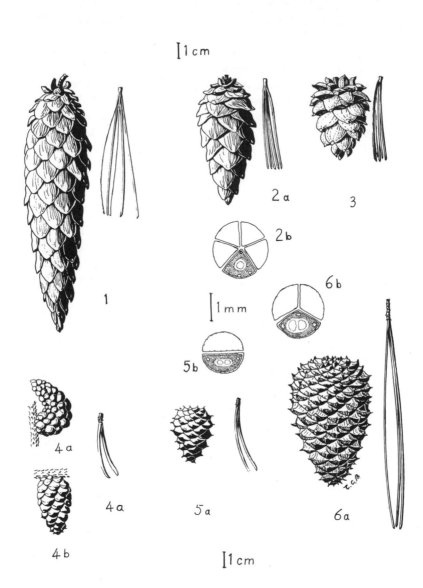

1 cm

2 a

3

2 b

1

1 mm

6 b

5 b

4 a

4 a

5 a

6 a

4 b

1 cm

Plate 8 Comparison of Pine Cones and Foliage. Figures: (1) *Pinus monticola;* (2) *P. flexilis;* (3) *P. albicaulis;* (4a) *P. banksiana;* (4b) *P. × murraybanksiana;* (5) *P. contorta;* (6) *P. ponderosa.* 2b, 5b, 6b, cross sections of leaves and veins.

Cupressaceae   Cypress Family

Columnar trees or variously shrubby. Leaves opposite or whorled; needle-like on juvenile growth; scalelike and overlapping, or sometimes short and sharply needlelike, on adult growth. Pollen cones minute, their stamens with 2-6 pollen sacs. Seed cones small, dry or berrylike, with opposite, often thick scales that may become fleshy, with 2-5 erect seeds attached to each scale. Seedlings usually with 2 (or sometimes more) cotyledons.

Key to the Genera of Cupressaceae

1a Leaves needlelike, in whorls of 3. Fruit berrylike . . . . *Juniperus*
1b Leaves of mature branches scalelike, opposite.
   2a Branches branching in 2 ranks, forming flat sprays. Cone dry when ripe.
      3a All leaves on adult trees tightly appressed. Cones ovoid, with overlapping scales . . . . *Thuja*
      3b Leaves on thrifty branchlets spreading at tips. Cones globular, with umbrella-shaped scales . . . . *Chamaecyparis*
   2b Branches branching in 4 ranks, forming thick sprays. Cones berry-like when ripe . . . . *Juniperus*

Thuja   Arborvitae

Medium to large trees of columnar form with thin bark, becoming fibrous on old trees. Branching 2-ranked, the branchlets forming flat sprays. Leaves of mature trees small, scalelike, closely appressed and overlapping in opposite pairs, the lateral pairs larger than the partly concealed facial pairs. Cones monoecious, terminal on lateral branchlets; staminate cones minute; ovulate cones erect, solitary or in clusters, each cone with 4-6 pairs of opposite, thin, overlapping scales without distinct bracts; the 2-4 fertile scales each bearing two seeds with paired lateral wings. Seedlings with 2 cotyledons and linear, pointed, spreading leaves; scale-leafed shoots arise in the axils of these linear leaves.

Thuja plicata Donn *ex* D. Don in Lambert   **Western Red-cedar**
Giant Arborvitae

Large tree with a steeply conical crown, up to 70 m tall and 2-4 m in diameter; the trunk markedly tapering, especially at the fluted and buttressed bases of old trees. Bark thin, reddish brown, shallowly furrowed, fibrous and shredding on old trees. **Leaves** scalelike, rounded, light yellowish green; the rounded, incurved scale tips giving a branchlet a smoother feel than that of *Chamaecyparis*. **Cones** 1-1.5 cm long, ellipsoidal; maturing in one season, falling the next. Seeds dark brown, with thin, pale wings that are separate at the seed apex. Seedlings with 2 cotyledons 6 mm long, and linear leaves 5-8 mm long. Plate 9, figure 1.

Range: Southeastern Alaska to northwestern California, from the coast to the Cascade and Coast ranges and, inland, in the Columbia Forest Region and eastward to Montana. In British Columbia, on the coast from sea level up to 1,300 m altitude; in the interior, northward to lat. 54°30'N and from the Okanagan Valley eastward to the Rocky Mountains, extending up to 1,370 metres in the Selkirk Range. A characteristic and important tree of alluvial flats. A columnar, compact form of this species has been found on the Kilgard Indian Reserve near Abbotsford and in the Kootenay River valley between Nelson and Castlegar (25).

Chamaecyparis

Trees of aspect similar to that of *Thuja*. Cones monoccious. Ovulate cones globular, with 2-3 pairs of scales that are round and centrally stalked; each scale with a prominent central point (umbo) and bearing 2-4 seeds. Each seed with a surrounding wing. Cones mature in one season at low altitudes, but in their second year at high altitudes. Seedlings with 2 cotyledons.

Chamaecyparis nootkatensis   **Yellow-cedar**
(D. Don in Lambert) Spach   **Yellow Cypress**

Tree up to 30 m tall and up to 90 cm in diameter, sometimes with more than one trunk; branches rather widely spaced, the secondary branches often hanging down, giving the tree a "weeping" aspect. Bark thin, greyish to chocolate-brown, ultimately shredding into flat, ribbonlike ridges.

Leaves angular rather than rounded, with slightly divergent tips, especially on vigorous terminal shoots, giving a branchlet a harsher feel than that of *Thuja;* foliage dark green on both sides. **Cones** globose, up to 1.2 cm in diameter; scales thick, hard and brown at maturity; tipped with sharp, sometimes reflexed, umbos. Seedlings with 2 cotyledons about 15 mm long. Plate 9, figure 2.

Range: From southern Alaska to northern California along the coast and western slopes of the Coast and Cascade ranges; mainly in the Coastal Subalpine Forest Zone, with a few new records further inland. In British Columbia, on the coastal islands mostly between 600 and 900 m altitude; extending down to sea level, where it prefers swampy habitats, from Vancouver Island northward. A small, isolated area occurs at around 1,550 m altitude near Slocan Lake, in the West Kootenay region.

Juniperus   Junipers

Evergreen trees or shrubs with thin, stringy bark. Leaves either short, minute, scalelike, sharp-pointed, arranged in opposite pairs or sometimes in threes, with bases extending down the stem, or with branchlets bearing much longer needlelike leaves that stand out loosely in whorls of 3. Cones dioecious, terminal or in the axils of leaves. Ovulate cones globular, berrylike by fusion of the fleshy scales. Seeds 1-3. Cotyledons 2, needlelike.

Key to the Species of *Juniperus*
1a Leaves normally opposite, usually scalelike; cones terminal on short lateral branchlets.
    2a Trees, or sometimes ascending shrubs . . . . *J. scopulorum*
    2b Prostrate shrubs . . . . *J. horizontalis*
1b Leaves in whorls of 3, all needlelike; cones axillary . . . *J. communis*

Juniperus scopulorum Sargent   Rocky Mountain Juniper

A small tree, up to 20 m tall and 40 cm in trunk diameter, or a big ascending shrub of shaggy appearance. Branches ascending, often from near the base in exposed situations. **Leaves** mostly opposite, scalelike, on slender, 4-sided branches, or sometimes shortly needlelike on young vigorous shoots; light green and glaucous or dark green. **Cones** terminal on short branchlets. **Fruiting Cones** berrylike, 6-8 mm long, on short recurved branchlets; fleshy, greenish to purplish, with a blue-grey bloom; sweetish, ripening the second season. Seeds 1 or 2, acute, grooved and angled, 4-5 mm long, germinating in the third season. Plate 9, figure 3. Photo 16.

Range: On dry or rocky soils, sometimes on riverbanks or lakeshores; from the coast across British Columbia to the Rocky mountains and over much of the western United States. Northward on southern exposures to the Skeena River near Hazelton, with an isolated area around Telegraph Creek, on the Stikine River (lat. 58°N).

Note: Under heavy browsing, this species may be reduced to a broad, ascending shrub, which may be mistaken for *J. horizontalis.*
 An exceptional tree at Deep Cove, Vancouver Island (lat. 48°41'N, long. 123°38'W), was reported as 24 m tall and 90 cm in diameter (Freeman F. King).

Juniperus × fassettii Boivin

This natural hybrid between *J. scopulorum* and *J. horizontalis* is sometimes seen where these species occur together. It forms a wide shrub, 1-3 m tall and often wider than high, with spreading and obliquely ascending branches and no distinguishable central trunk or only a very short one. It has been found at Dutch Creek in the Rocky Mountain Trench, Alexis Creek in the Cariboo Region and Day's Ranch in the Stikine Valley.

Juniperus horizontalis Moench

Creeping Juniper
Horizontal Juniper

Trailing shrub with long branches and numerous short branchlets. **Leaves** typically scalelike and opposite, but shortly needlelike and transitional leaf forms often occur on vigorous long shoots, acute, steel-blue and glandular on the back. **Fruit** blue, bloomy, usually on recurved stalk, maturing in one year; seeds 1-4, usually 2. Plate 9, figure 4.

Range: Boreal-transcontinental and southward along the Rocky Mountains to Colorado. Widespread in central and northern British Columbia east of the Coast range, in open woods and balds, at low to middle elevations.

Juniperus communis L. var. *saxatilis* Pallas   Common Juniper
(*J. sibirica* Burgsdorf and
J. communis var. *montana* Aiton)

Spreading or obliquely ascending shrub, prickly to handle. **Leaves** in whorls of 3, all needlelike, 8-15 mm long, stiff, sharply pointed, grooved and glaucous on the upper surface, the attachment not elongated. **Cones** very short-stalked, in the axils of the leaves. **Fruiting cone** berrylike, greenish to dark blue and slightly bloomy, 5-9 mm in diameter; seeds 1-3, ovoid, acute, angled, about 3 mm long. Plate 9, figure 5.

Range: *J. communis* is widespread over the Northern Hemisphere. Var. *saxatilis* is transcontinental in North America and occurs in Siberia. It is found in open and semi-open sites on dry, stony soils from sea level to alpine elevations in the mountains. In Europe the typical form of the species is a small tree.

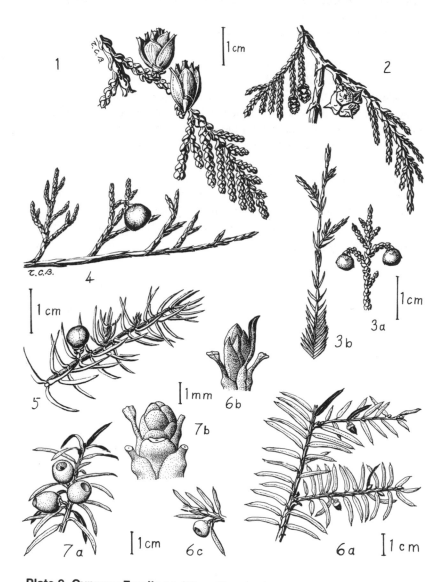

Plate 9 Cypress Family and Yew. Figures: (1) *Thuja plicata;*
(2) *Chamaecyparis nootkatensis;* (3) *Juniperus scopulorum:* 3b, juvenile
foliage; (4) *J. horizontalis;* (5) *J. communis;* (6) *Taxus brevifolia:* 6c, fruit;
(7) *T. baccata.* 6b, 7b, terminal, vegetative buds with surrounding leaves
removed.

Taxaceae   Yew Family

Evergreen, dioecious trees or shrubs with alternate, linear, flat leaves acute-tipped and tapered to short petiolar bases. Cones minute, in the axils of leaves and with minute scales. Pollen cones globular, of 3-12 stamens; the stamens umbrella-shaped, with 2-9 pendent pollen sacs. Seed cones with few cone scales and one terminal, erect, protruding ovule provided with an aril (outgrowth of the ovule base). Seed fully exposed or enclosed in its aril. Seedlings with 2 cotyledons. A small family of 3 or 4 widely dispersed genera.

Taxus   Yews

Dioecious evergreen trees or shrubs with scaly bark, green twigs and strong, elastic, dense, reddish wood. Leaves alternate, linear, with very short petioles. Pollen cones minute, globular; the stamens bearing usually 6 pollen sacs. Seed cones consisting of a few pale scales and a terminal, exposed, green ovule bearing a collarlike green aril around its base. The fruit consists of the dark greenish seed surrounded by a red, fleshy, cup-shaped aril. Young seedlings resemble the adult in foliage, with 2 short, pointed cotyledons. This genus consists of about 8 species dispersed around the Northern Hemisphere.

Key to the Species of *Taxus*
1a Branches spreading. Vegetative buds with acute, keeled, loose outer scales. Ripe aril at least as wide as long, the cup shallow . . *T. brevifolia*
1b Branches ascending. Vegetative buds with rounded, not keeled, closely appressed outer scales. Ripe aril barrel-shaped, rather longer than wide, forming a deep cup . . . . *T. baccata*

Taxus brevifolia **Nuttall**   **Pacific Yew**
                                                  Western Yew

Typically a small understorey tree; commonly 10 m tall and 30 cm in trunk diameter, rarely up to 20 m tall and 60 cm in trunk diameter, with straight, fluted trunk; but often a big shrub in the interior. The crown is often irregular, with slender, horizontal branches and rather drooping branchlets. Bark thin, reddish, with thin papery scales. Vegetative buds with acute, keeled, rather loosely overlapping outer scales. **Leaves** appearing 2-ranked, flat, linear to slightly tapering, ridged over the midrib above, with an abruptly acute to acuminate apex and a short petiole. **Fruit** a dark greenish seed 4-6.5 mm long, inserted in a rather shallow cuplike red, fleshy aril, 8-9 mm long, normally wider than long. Seen from a distance, this species presents a rather open crown of dull, greyish green foliage, in contrast to the intensely green, dense crown of the European Yew. Plate 9, figure 6.

Range: Alaska to California and Montana. Widespread in the Coast Forest and Columbia Forest regions in British Columbia (53).

Note: The fruit ripens in September and is often eaten by birds for its sweet aril, thus providing for the dispersal of the seed. Beware, however, that the seed is poisonous to humans.

Taxus baccata **L.**   **European Yew**

Small tree or shrub, 10 to rarely 75 m tall, with ascending or erect branches. Twigs bright green; the vegetative buds with closely appressed, rounded, unkeeled scales. **Leaves** spreading all around the twigs, seldom appearing 2-ranked; the foliage dense and deep green. **Fruit** a red, fleshy, slightly elongate, barrel-shaped aril, 10-13 mm long, forming a deep cup around the 4-6 mm long seed. Plate 9, figure 7.

Range: Native of Europe. Commonly planted in many horticultural forms (37) as an ornamental subject, especially on the coast. Occasionally escaping on southern Vancouver Island, it may be mistaken for the native *Taxus brevifolia.*

Salicaceae   Willow Family

Deciduous, dioecious shrubs and trees, with bitter bark. Leaves usually alternate, simple and petioled, with or without stipules. Flowers in separate staminate and pistillate catkins that are lateral on branches or terminal on short lateral branchlets. Each flower is subtended by a minute bract. Fruits are small capsules; the 2 or 3 valves separate to shed the tiny seeds, each of which bears a circle of fine silky hairs to facilitate dissemination by the wind (5, 7).

Key to the Genera of Salicaceae
1a   Buds covered by a single scale; leaves usually narrow; bracts of
catkins entire .... *Salix*
1b   Buds with several scales; leaves usually broad; bracts of catkins lobed
.... *Populus*

Populus   Poplars

Trees with more or less resinous buds covered by many overlapping scales, the lowest scale directly above the leaf scar. Twigs with 5-angled pith. Leaves alternate, broad; petioles long; stipules early deciduous. Catkins pendent, emerging before the leaves; flowers pollinated by wind, dioecious, borne in a cup-shaped calyx, subtended by a lobed and fringed bract; stamens 6-12 or numerous; styles short; stigmas 2-4; capsules have 2-4 valves; seeds small and tufted.

Keys to the Species of *Populus*
1a   Twig and underside of leaf covered with dense, matted white hairs;
leaf margin sinuate to palmately lobed .... *P. alba*
1b   Twig glabrous or thinly puberulent; leaf similar when mature and
with toothed margin.
    2a   Petiole compressed in a plane perpendicular to leaf blade.
       3a   Leaf blade nearly circular to broadly ovate.
          4a   Leaf margin with many fine teeth .... *P. tremuloides*
          4b   Leaf margin coarsely toothed, with 12 or fewer teeth
          .... *P. grandidentata*
       3b   Leaf blade broadly triangular to rhombic.

5a Branches spreading; leaf triangular, coarsely toothed except
for the entire, attenuate apex ...., . *P. deltoides*
5b Branches ascending to erect; leaf broadly rhombic, finely
toothed to the shortly acuminate apex .... *P. nigra* var. *italica*
2b Petiole cylindrical and grooved on top; leaf greyish and resin-
stained beneath .... *P. balsamifera*
6a Fruit ovoid, pointed, glabrous, 2-valved; stamens 12-30
.... *P. balsamifera* subsp. *balsamifera*
6b Fruit globular, hairy, 3-valved; stamens 30-60
.... *P. balsamifera* subsp. *trichocarpa*

Populus balsamifera L. subsp. *balsamifera*　　　Balsam Poplar

Tree similar to Black Cottonwood and completely intergrading with it
across central and northern British Columbia and Alberta. **Leaves** on average rather narrower and more rounded at base; sucker shoots circular in
cross section, not angled as it often is in Black Cottonwood. **Flowers:**
male flowers with 12-30 stamens; female flowers with 2-valved ovary.
Fruit a 2-valved ovoid glabrous capsule. Plate 10, figure 1.

Range: Northern British Columbia to Alaska and to the Atlantic coast,
mainly in the Boreal Forest Region. The common "cottonwood" on the
northern plains east of the Rocky Mountains; in British Columbia, found
from the Peace River westward to Stuart Lake, Stikine River and Atlin.

Populus balsamifera L.　　　(Northern) Black Cottonwood
subsp. *trichocarpa* (Torrey & Gray) Brayshaw
(*P. trichocarpa* Torrey & Gray)

An important tree in British Columbia and the largest native broadleaf, it
is commonly 40 m tall and 1-2 m in trunk diameter. Bark grey, changing
from smooth to deeply furrowed, with hard, narrow, flat-topped ridges.
Buds often 20 mm long, the terminal stouter than laterals; laterals sharp-
pointed, divergent or appressed, often very resinous and usually fragrant.
Leaves 75-200 mm or more long, broadly ovate with rounded to cordate
base; thick, glabrous or rusty glaucous beneath; margins finely crenate-
serrate to almost entire; petioles round, half as long as the blade. **Catkins**
long; bracts fringed, deciduous, slightly pubescent; stamens 40-60; stigmas 3; capsule nearly globose, pubescent, nearly sessile, 3-valved. Plate
10, figure 2. Photo 17.

Range: Alaska to Alberta, Montana and California, at low and high altitudes along streams and floodplains. Throughout British Columbia; partly replaced east of the Rockies in northeastern British Columbia by Balsam Poplar.

Populus tremuloides Michaux   Trembling Aspen

Medium-sized, slender tree, from an extensive, shallow, freely suckering root system, forming groves. Bark initially smooth, white in the interior to greenish grey on the coast, becoming furrowed and dark grey with age. **Leaves** circular to broadly ovate, thin, 15-75 mm across; margins finely crenate-serrate; apex shortly acuminate, base rounded, truncate or subcordate; surfaces glabrous, locally tomentose at first, lighter and duller below; petioles slender, as long as leaf blade, laterally flat. Leaves of suckers are often very large and can be mistaken for those of Black Cottonwood. **Flowers** in hanging catkins; bracts incised and fringed with long hairs; stigmas 2; stamens 6-12, usually 8; **Fruit** a slenderly conical 2-valved capsule. Plate 10, figure 3.

Key to the Varieties of *P. tremuloides*
1a Smooth bark bright white; twigs and buds glabrous; leaves thin, 25-45 mm across; interior and eastward . . . . **var.** *tremuloides*
1b Smooth bark greenish grey; twigs and buds finely puberulent; leaves larger, rather thick; coastal . . . . **var.** *vancouveriana*

Range: Transcontinental as a species and found over most of North America. Var. *tremuloides* is common at moderate altitudes across the interior of British Columbia, especially on finer-textured soils in moist areas and dominating burnt sites quickly by suckering until overtaken by young conifers reinvading by seed (Photo 18). Var. *vancouveriana* (Trelease) Sargent occurs in scattered localities on Vancouver Island and the southern mainland coast, in open sites.

Populus deltoides Bartram   Plains Cottonwood
ex Marshall var. *occidentalis* Rydberg

Tree to 25 m tall with a broad head of spreading branches; bark grey, becoming darker and deeply fissured on the lower trunk; branchlets yellowish, angular when growing vigorously; buds gummy, finely tomentose. **Leaves** on flat petioles, the blade broadly triangular, coarsely toothed except for the entire acuminate tip, green on both sides. **Flowers:** staminate flowers with 40-60 stamens; pistillate flowers with 3-carpelled ovary. **Fruit** ellipsoid, 3-valved. Plate 10, figure 5.

Range: A common riverbank tree of the Great Plains region, this species has been introduced into British Columbia and now colonizes riverbanks and shorelines at Creston and Osoyoos.

Populus nigra L. var. *italica* Duroi   **Lombardy Poplar**
(*P. italica* [Duroi] Moench)

Stiffly and narrowly erect tree with strongly ascending to vertical branches and a flaring base. Bark initially smooth and pale grey, becoming light yellowish brown, rough and furrowed. **Leaves** broadly rhombic to nearly triangular, on compressed petioles; the apex shortly acuminate; the margins rather finely toothed; glabrous. Pistillate trees rare or absent. Plate 10, figure 4.

Range: Believed to be a native of eastern Europe and western Asia. Abundantly planted all over Europe and much of North America; propagated by cuttings; persisting at the sites of planting.

Populus grandidentata Michaux   **Large-toothed Aspen**

Erect slender tree arising from a spreading and suckering root system. Bark initially smooth and white, becoming dark grey and fissured. Twigs brownish grey, downy when young. **Leaves** ovate to nearly circular, on long, compressed petioles; the margin sinuately toothed with 4-10 coarse teeth; the petiole and leaf surface covered with white down while unfolding, later becoming glabrous or nearly so. **Flowers** and **Fruit** as in *P. tremuloides*. Plate 10, figure 6.

Range: From southeastern Manitoba to Nova Scotia and southward to North Carolina. Found mostly on dry, coarse-sandy and gravelly soils. Rarely introduced and established in British Columbia, as at Hope.

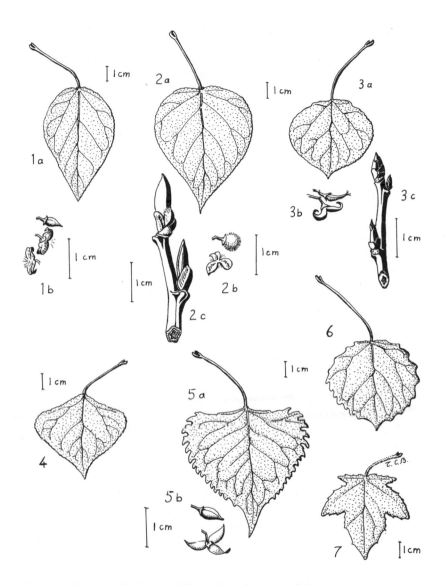

Plate 10 Poplars, Native and Introduced. Leaves (a) to a common scale; capsules (b) and twigs (c) to a common scale. Figures: (1) *Populus balsamifera* subsp. *balsamifera;* (2) *P. balsamifera* subsp. *trichocarpa;* (3) *P. tremuloides;* (4) *P. nigra;* (5) *P. deltoides;* (6) *P. grandidentata;* (7) *P. alba.*

Populus alba L.   White Poplar

Wide-crowned tree up to 75 m tall, suckering freely. Twigs, buds and petioles white-tomentose, at least when young. **Leaves** ovate to pentagonal, 3-10 cm long, shallowly palmately lobed and coarsely dentate; tomentose above when young, permanently white-tomentose beneath. **Catkins:** staminate catkins up to 6 cm long, the flower with 6-10 stamens; pistillate catkins up to 6 cm long; the bracts bluntly toothed and ciliate, the stigmas linear. Plate 10, figure 7.

Range: This relative of the aspens is native to southern Europe and Asia; planted in this country as an ornamental tree and persistent about the sites of its planting, probably mainly by suckering. It can hybridize with *P. grandidentata.*

Salix   Willows

Shrubs, sometimes prostrate, or trees, with each bud enclosed in a single bud scale whose margins are joined together on the back of the bud in all our species except *Salix amygdaloides.* Terminal bud absent. Leaves alternate, usually narrow, unlobed, on short petioles, with or without stipules. Staminate and pistillate flowers on separate plants, in catkins. Catkins erect or spreading, not pendulous; usually from the lower axillary buds on the previous season's twig; sessile or on short, leafy branchlets; expanding before (precocious), with (coetaneous) or after (serotinous) the leaves. Bracts of the flowers small, scalelike, usually entire. Staminate flowers with 1-5 (usually 2) stamens in our species (up to 12 in some exotic species). The pistillate ("female") flower has a single 1-chambered ovary, sessile or stiped, in the axil of the bract, accompanied by one or two nectar glands. Style none or present, sometimes partly divided. Stigmas 2, often forked. The ripened ovary forms a capsule that splits down 2 sides from the tip, freeing numbers of small seeds, each seed surrounded by long white hairs, which facilitate dispersal by wind (1). Willow flowers may be pollinated by insects or, in their absence, by the wind.

Hybrids have been reported between many species of *Salix* (1, 5). They are not specifically dealt with here, except for one (*S.* × *rubens*) that is spreading into the wild locally (7).

Key to the Species of *Salix*

This key depends on specimens with well-developed pistillate flowers or fruit and is based on the key in *Catkin-Bearing Plants of British Columbia* (7), revised from its first edition (5) in view of the more up-to-date treatment by Argus in *The Vascular Plants of British Columbia* (16). Sizes refer to those of mature organs. Because of the large number of species in *Salix*, each species is numbered to make its description easier to find.

1a Low-growing dwarf shrubs of alpine tundra, with prostrate or subterranean rooting branches, often mat-forming or cushion-forming. Catkins coetaneous or serotinous, on leafy branches.

  2a Leaf veins usually deeply impressed above and raised-reticulate beneath. Leaf glaucous beneath. Margin revolute. Glands 2 per flower, or if only 1, bifid. Capsule usually hairy.

    3a Leaves broadly elliptic to orbicular, rounded to subcordate at base. Veins usually deeply impressed above. Catkins elongate, many-flowered .... **1. *S. reticulata* subsp. *reticulata***

    3b Leaves narrowly elliptic to oblanceolate, tapering to base. Veins relatively shallowly impressed above. Catkins short, usually few-flowered (up to 25 flowers) .... **1. *S. reticulata* subsp. *nivalis***

  2b Leaf veins not or scarcely impressed above. Gland 1 per flower.

    4a Leaves green on both sides.

      5a Branches covered with persistent, skeletonized, old dead leaves. Leaf ciliate-margined. Capsule thinly pubescent with wavy lustrous hairs. Stipe longer than gland

        .... **6. *S. phlebophylla***

      5b Branches not covered with old skeletonized leaves. If leaves sometimes persistent, not skeletonized.

        6a Leaves 1 cm or less long, ciliate. Veins raised on both surfaces. Capsule glabrous or almost so ..... **7. *S. rotundifolia***

        6b Leaves over 1 cm long and usually not ciliate at maturity. Veins not raised on upper surface. Capsule pubescent

        .... **3. *S. polaris***

    4b Leaves glaucous beneath.

      7a Leaves lanceolate, 3 or more times as long as wide, acute at ends, 1-3 cm long, with closely ascending lateral veins. Catkins 1-2 cm long. Capsule usually hairy, up to 5 mm long

        .... **4. *S. cascadensis***

      7b Leaves elliptic, varying to oblanceolate or orbicular, 3 or less times as long as wide.

8a Capsules hairy. Leaves obovate to oblanceolate, 2-8 cm long, greyish green. Catkins 2-8 cm long .... **2. *S. arctica***
8b Capsules glabrous. Leaves elliptic to orbicular, 1-2.5 cm long, deep green above .... **5. *S. stolonifera***
1b Normally upstanding shrubs 20 cm tall or taller, or small trees; or if prostrate or depressed in alpine situations, the branches not rooting.
9a Colonial riverbank or beach shrubs, with erect shoots from long horizontal roots. Leaves pale greyish green, the midvein often raised above. Petioles 3 mm or less long. Catkins on leafy branchlets.
10a Leaves oblong-elliptic, sessile, 1.5-4 times as long as wide, glabrous and somewhat fleshy, with flat, dark purplish midvein and no stipules. Catkin 2 cm or less long. Up to 30 cm tall. Far northwestern, on sandbars .... **10. *S. setchelliana***
10b Leaf linear to lanceolate, (2.5-)5-20 times as long as wide, the midvein tending to be raised on both surfaces. Catkins late, often more than 1, on well-developed leafy branchlets. Bract pale yellowish, usually deciduous early. Often more than one order of branching in one season's growth. 1-3 m tall. Widespread. The Sandbar Willows
11a Leaves rather densely greyish-silky-hairy, 5-8 times as long as wide. Style present but short. Stigmas slender. Coastal .... **9. *S. sessilifolia***
11b Leaves thinly silky-hairy to glabrous, 5-20 times as long as wide. Style none. Stigmas stout. Widespread .... **8. *S. exigua***
9b Noncolonial shrubs or trees with single root-crowns. Leaf midvein usually not raised above.
12a Leaves lanceolate or sometimes oblanceolate, usually acuminate at apex and rounded to acute at base, finely and regularly serrulate, usually shiny above even when hairy. Twigs and buds usually yellowish. Tall shrubs or trees. Capsules glabrous. Bracts of pistillate flowers usually deciduous early, except in *S. prolixa.*
13a Leaves with 2 to several prominent, raised glands at junction of petiole with blade. Stipules often conspicuous. Staminate flower with 2 glands and 5 stamens.
14a Leaves lanceolate, broadest at mid-length or below, acuminate to attenuate at apex. Catkins usually 5 cm or more long; widespread .... **13. *S. lucida***
14b Leaves oblanceolate, acute at apex. Catkins usually 3.5 cm or less long. East of Rocky Mountains .... **14. *S. serissima***

13b Petiolar glands, if present, minute, inconspicuous and more or less flush with the adjacent surface.

15a Stipules usually conspicuous, at least on vigorous shoots. Leaves rounded to cordate at base. Bracts dark, persistent. Stamens 2 per flower. Upper side of petiole more hairy than elsewhere . . . . **22. *S. prolixa***

15b Stipules minute, deciduous early. Leaves tapering or acute at base. Bracts usually deciduous early. Upper side of petiole glabrous, or at least not more hairy than elsewhere.

16a Bud scale margins overlapping on side of bud adjacent to twig. Leaves glabrous on both surfaces. Petioles 10-25 mm long. Stamens 5 . . . . **11. *S. amygdaloides***

16b Bud scale margins united. Leaves at least thinly silky-hairy. Petioles stiff, 4-10 mm long. Stamens 2. Introduced tree or shrub . . . . **12. *S. alba***

12b Leaves various: not with the above combination of characteristics. Twigs and buds usually not yellow. Bracts normally persistent.

17a Underside of mature leaves concealed by densely matted white hairs.

18a Leaves rhombic, elliptic, obovate or broadly lanceolate; 2-4 times as long as wide . . . . **29. *S. alaxensis***

18b Leaves linear to oblong, more or less parallel-sided, 5-8 times as long as wide . . . . **30. *S. candida***

17b Underside of mature leaf glabrous, or if hairy the hairs not matted and not concealing the surface.

19a Leaves broadly elliptic, wrinkled above with deeply impressed veins that are raised and reticulate on the whitened lower surface, which bears abundant long, silky hairs. Floral glands commonly 2, one of them very much reduced
. . . . **45. *S. vestita***

19b Leaves otherwise.

20a Mature leaves with persistent band of short hairs on upper side of midvein and petiole. Capsules glabrous.

21a Leaves oblong-lanceolate to obovate, 3.5 or more times as long as wide. Stipules often large, orbicular or broadly ovate. Stipe usually 2 mm or more long
. . . . **22. *S. prolixa***

21b Leaves elliptic to obovate, 3 or less times as long as wide. Stipe usually 2 mm or less long.

22a Catkins arising from ultimate and penultimate buds on twig, sessile. Leaves green beneath. Style 1.5-2 mm long . . . . **43. *S. tweedyi***

22b Catkins from lower lateral buds, with leafy
branches from ultimate and penultimate buds. Leaves
glaucous beneath.
  23a Mature leaves equally hairy on both sides, or
  glabrous. Catkins serotinous, on leafy branchlets
  .... **20. S. commutata**
  23b Mature leaves glabrous beneath, glabrescent or
  with persistent hairs above; if glabrous above,
  catkins precocious.
  23c Mature leaves more hairy beneath than above.
  Catkins sessile, large .... **28. S. hookeriana**
    24a Stipules persistent into second year or later.
    Catkins almost sessile .... **44. S. lanata**
    24b Stipules deciduous during first season.
      25a Leaves glabrous on both sides when fully
      mature (except midvein above). Fertile
      branchlet less than 1 cm long, with leaves up
      to 1.5 cm long ..... **21. S. pseudomonticola**
      25b Leaves with some hair persistent above.
      Fertile branchlet 1-3 cm long with larger
      leaves .... **19. S. barclayi**
20b Mature leaves not differently hairy on upper side of midvein from rest of leaf surface, or if so, the capsule usually hairy.
  26a Mature leaves satiny beneath with appressed, parallel,
  lustrous hairs.
    27a Twigs distinctly pruinose, with a bluish grey bloom,
    glabrous .... **32. S. drummondiana**
    27b Twigs not pruinose.
      28a Twigs glabrous or soon becoming so. Leaves nar-
      rowly lanceolate.
        29a Twigs dark reddish and shiny. Bracts blackish
        at least toward tip. Stipe 1.2 mm or less long.
        Stamens glabrous. Style 0.3-0.5 mm long.
        Flowering coetaneous .... **35. S. arbusculoides**
        29b Twigs yellowish or light reddish brown, eventu-
        ally blackening. Bracts yellowish or brown. Stipe
        1.5 mm or more long. Stamens hairy at base
        .... **36. S. petiolaris**
      28b Twigs hairy at least to end of first season's growth.
      Leaves obovate to broadly lanceolate.
        30a Leaves usually strongly satiny beneath. Stipe

less than 1 mm long. Stamen 1 per flower. Bracts dark . . . . **31. *S. sitchensis***
30b Leaves inconspicuously satiny beneath. Stipe 2-5 mm long. Stamens 2 per flower. Bracts pale, yellowish . . . . **25. *S. bebbiana***
26b Mature leaves glabrous beneath, or if hairy, the hairs not appressed and parallel, not producing a satiny lustre.
  31a Leaves green and glabrous beneath at maturity.
    32a Leaves entire. Petioles usually 3 mm or less long
    . . . . **39. *S. pedicellaris***
    32b Leaves serrulate-margined. Petioles more than 3 mm long.
      33a Twigs dark brown to purplish. Bracts 3-5 mm long, yellow to pale brown, hairy at base, glabrous at tip. Capsules 9-12 mm long, white-hoary. Stamens hairy at base
      . . . . **37. *S. maccalliana***
      33b Twigs greenish to yellowish brown. Bracts 1 mm or less long, brown to black, hairy. Capsules 3-6 mm long, glabrous. Stamens glabrous.
        34a Leaves acute to rounded at tip. Stipules blunt, 1-5 mm long or absent
        . . . . **40. *S. myrtillifolia***
        34b Leaves abruptly short-acuminate at tip. Stipules sharply acute, 5-12 mm long on vig-orous shoots . . . . **41. *S. boothii***
  31b Leaves glaucous beneath at maturity
    35a Mature leaves glabrous beneath.
      36a Bracts yellowish or pale brown throughout.
        37a Bracts and capsules short-hairy. Stipe 2-5 mm long . . . . **25. *S. bebbiana***
        37b Bracts and capsules glabrous.
          38a Petioles 3 mm or less long. Stipe 2-4 mm long. Style none. Stigmas stout
          . . . . **39. *S. pedicellaris***
          38b Petioles 5-10 mm long. Stipe 0.4-1.2 mm long. Style 0.6-1 mm long. Stigmas slender
          . . . . **18. *S. raupii***
      36b Bracts at least partly dark brown to black, usually hairy.
        39a Bracts dark at tip, pale toward base.

40a Twigs purplish. Leaves rhombic-lanceolate, 3-5
times as long as wide. Capsules hairy. Catkins precocious, more or less sessile .... **38.** *S. planifolia*
40b Twigs reddish brown. Leaves elliptic to lanceolate,
2-3 times as long as wide. Capsules glabrous. Catkins
on leafy branchlets .... **24.** *S. hastata*
39b Bracts dark throughout.
41a Mature twigs normally greyish-hairy, although
sometimes sparsely so; or, if twig glabrous, leaf distinctly obovate, obtuse to rounded at apex and tapering
to base. Stipe 0.5-1.5 mm long. Style 0.5 mm or less
long .... **27.** *S. scouleriana*
41b Twigs glabrous or becoming so at maturity.
42a Leaves tapering to base. Capsules hairy.
Flowering precocious.
43a Leaves 2-4 times as long as wide, usually
dull above, coarsely crenate to entire. Stipe 1.4-
2.4 mm long. Stamens short-hairy at base
.... **26.** *S. discolor*
43b Leaves 3-5 times as long as wide, shiny
above, entire to serrulate. Stipe 1 mm or less
long. Stamens glabrous ...... **38.** *S. planifolia*
42b Leaves rounded to cordate at base. Capsules
glabrous.
44a Stipe 1.5 mm or less long. Flowering precocious .... **21.** *S. pseudomonticola*
44b Stipe 2-4 mm long. Flowering coetaneous.
Foliage often with balsamic aroma
.... **23.** *S. pyrifolia*
35b Mature leaves with at least some hairiness beneath.
45a Foliage pale or greyish green above, even when not
densely hairy. Capsules usually greyish-hairy.
46a Catkins sessile, often at end of branch. Buds and
twigs often oily or sticky. Bracts black. Leaves ascending,
greyish-hairy, often aromatic .... **42.** *S. barrattiana*
46b Catkins on short, leafy branchlets, with sterile leafy
branches arising more distally to them. Buds and twigs
"dry." Foliage not aromatic. Bracts usually yellowish.
47a Petioles 4-10 mm long, longer than the subtended
bud. Catkins 3-5 cm long, densely flowered
.... **15.** *S. glauca*

47b Petioles 1-4 mm long, often shorter than the subtended bud. Catkins shorter in *S. glauca or* loosely flowered if as long .... **16. S. brachycarpa**
45b Foliage deep or bright green above.
48a Leaves lanceolate, 3-8 times as long as wide, acuminate at apex.
49a Twigs purplish, glabrous at full growth. Catkins precocious, more or less sessile. Bracts acute to acuminate at apex. Stamens glabrous. Style 0.8-2.0 mm long .... **38. S. planifolia**
49b Twigs greenish to brown, darkening on drying, sometimes pruinose; hairy, although often thinly so, at maturity. Catkins coetaneous, on a short, small-leafed branchlet. Bracts rounded at apex. Stamens hairy at base.
50a Mature leaves thinly sericeous beneath. Pistillate catkins 1-2.5 cm long. Bracts more or less oblanceolate, at least twice as long as wide, pale to dark brown. Style 0.1-0.4 mm long .... **33. S. geyeriana**
50b Mature leaves glabrous or glabrescent beneath. Pistillate catkins 2-6 cm long. Bracts obovate, less than twice as long as wide, dark brown to black. Style 0.4-0.9 mm long .... **34. S. lemmonii**
48b Leaves obovate, elliptic or broadly oblanceolate, 2-3 times as long as wide, acute to rounded at apex.
51a Bracts pale brown or yellowish, usually shorter than the 2-5 mm long stipe. Stamens hairy .... **25. S. bebbiana**
51b Bracts medium brown to black, about as long as or longer than the stipe.
52a Low shrub up to 1 m tall. Leaves elliptic, 1-3 cm long. Catkins 1.5-2 cm long. Bracts brown. Northern .... **17. S. athabascensis**
52b Tall coarse shrubs or small trees. Leaves elliptic or obovate, 3-10 cm long. Catkins 2 cm long or longer. Bracts blackish.
53a Leaves distinctly obovate, tapering to base. Catkins 1-1.5 cm thick. Style 0.5 mm or less long .... **27. S. scouleriana**
53b Leaves elliptic to obovate, acute to rounded at base. Catkins 1.5-2.5 cm thick. Style 1-2 mm long .... **28. S. hookeriana**

1. *Salix reticulata* L.   Netted Willow
(including *S. nivalis* Hooker)

Prostrate shrub with smooth twigs purplish when young. **Leaves** broadly obovate to nearly circular or elliptic, entire or almost so; the surface wrinkled by the impressed reticulate veins above; very glaucous beneath, silky when young, becoming glabrescent; on long, reddish petioles. **Catkins** serotinous, terminal on branches, slender, reddish. Bracts 2 mm long, hairy, dark reddish at tips. Glands 2. Stamens 2, hairy at bases. **Capsules** 3-4 mm long, sessile, pubescent or glabrous. Styles short. Plate 11, figures 1-3. Photo 23

Key to the Subspecies of *S. reticulata* in British Columbia
1a Capsules and leaves glabrous. Endemic to the Queen Charlotte
    Islands . . . . **subsp.** *glabellicarpa* **Argus**
1b Capsules pubescent.
    2a Leaves 5-30 mm long, elliptic to oblanceolate, glabrous beneath,
        the veins not deeply impressed. Very dwarf plants of central and
        southern British Columbia
        . . . . **subsp.** *nivalis* **(Hooker) Love, Love & Kapoor**
    2b Leaves 10-65 mm long, broadly obovate to circular, silky beneath
        when young, with deeply impressed veins. Northern British
        Columbia . . . . **subsp.** *reticulata*

Range: Circumpolar and high alpine species, common on rocky well-drained or sometimes moist soils. Subsp. *reticulata* ranges across northern British Columbia and southward until it merges with subsp. *nivalis* between lat. 55°N and 56°N. Subsp. *nivalis* ranges southward on the mainland mountains (although absent from Vancouver Island) to New Mexico and California.

2. *Salix arctica* Pallas   Arctic Willow

Alpine species with trailing branches. **Leaves** entire, or obscurely toothed, pale green above, slightly glaucous beneath, obovate to oblong or elliptical, obtuse to rounded at apex, tapering to rounded at base, 2-6(-8) cm long, thin; petioles slender, yellow. **Catkins** 1-5(-9) cm long; bracts about 2 mm long, brown to black, long-hairy; glands oblong, 2.5-4 times as long as thick; stamens 2. **Capsules** 6-10 mm long, sessile or short-stiped, pubescent to glabrescent at maturity; style 1-2.5 mm long, sometimes partly divided. Plate 11, figures 4 and 5. Photo 24.

Key to the Subspecies of *S. arctica*

1a Leaves oblanceolate to obovate, at least twice as long as wide; bracts elliptic or pointed, brown or black; style 1-1.5 mm long
.... **subsp. *torulosa***

1b Leaves broadly obovate to nearly circular, 1.25-2 times as long as wide; bracts rounded to circular, black or black-tipped.

  2a Mature leaves obovate, two-thirds as wide as long, tapering to base, obtuse or rounded at apex, 3-6 cm long; capsule sessile; style about 1 mm long .... **subsp. *arctica***

  2b Mature leaves broader, circular or nearly so, rounded to subcordate at base; capsule short-stiped; style up to 2 mm long
.... **subsp. *crassijulis***

Range: The species is circumpolar, generally on rocky, well-drained tundra. Subsp. *arctica* is circumpolar on arctic tundra and ranges southward at alpine levels into British Columbia at Atlin. Subsp. *crassijulis* (Trautvetter) Skvortsov extends from Washington state through southern Alaska and the Aleutian Islands to Kamchatka, in mountains near the Pacific Coast; in British Columbia, mainly in the Coast range and on Vancouver Island; rarer eastward to the Rocky Mountains. Subsp. *torulosa* (Trautvetter) Hulten ranges from southern British Columbia through central and northern Alaska to Kamchatka and is widespread across British Columbia at alpine levels.

3. *Salix polaris* Wahlenberg   **Polar Willow**
subsp. *pseudopolaris* (Wahlenberg) Floderus

Prostrate, mat-forming shrub. **Leaves** rounded or obovate, acute or obtuse at apex, rounded at base, entire, glabrous or ciliate, green both sides. **Catkins** short, appearing with the leaves on leafy peduncles. Bracts up to 2 mm long, obovate, truncate, black, glabrous or thinly hairy. **Capsules** 5-7 mm long, nearly sessile, pubescent all over (in var. *pseudopolaris*) or only at the tip (in var. *glabrata* Hulten); styles 1 mm long or less. Plate 11, figure 6.

Range: From northern British Columbia and Northwest Territories westward across northern Eurasia. Across northern British Columbia at alpine elevations and southward in the Rocky Mountains to the Mount Robson area.

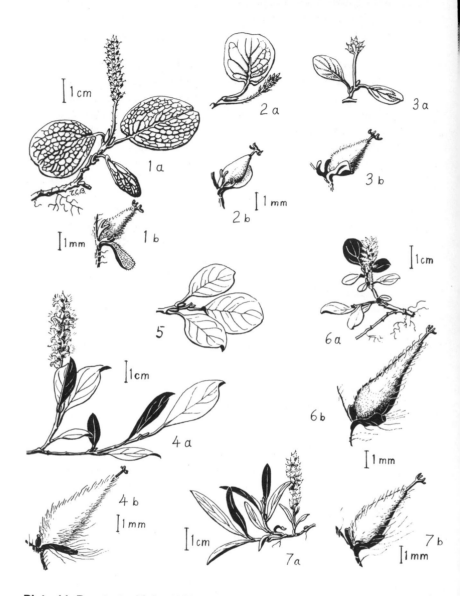

Plate 11 Prostrate Alpine Willows. Habit (foliage and catkins) (a) to a common scale; fruits (b) to a common scale. Figures: (1) *Salix reticulata* subsp. *reticulata;* (2) *S. reticulata* subsp. *glabellicarpa;* (3) *S. reticulata* subsp. *nivalis;* (4) *S. arctica* subsp. *torulosa;* (5) *S. arctica* subsp. *crassijulis;* (6) *S. polaris* subsp. *pseudopolaris;* (7) *S. cascadensis.*

4. *Salix cascadensis* Cockerell   Cascade Willow

Dwarf creeping alpine shrub. **Leaves** lanceolate, 1-2.5 cm long, 3-5 times as long as wide, with narrowly oblique lateral veins, glaucous beneath, cobwebby-hairy when young, becoming glabrous except for ciliate margins; on petioles 1-3 mm long. **Catkins** coetaneous, on short, leafy lateral branchlets, 1-2 cm long; bracts dark with hairs as long as bracts; stamens 2, sometimes united below; ovary and capsule usually tomentose, rarely glabrous (var. *thompsonii* Brayshaw). **Capsules** 4-5 mm long; style and stigmas 1-1.5 mm long. Plate 11, figure 7.

Range: British Columbia to Colorado and Utah on alpine slopes. In British Columbia, in the Cascade and southern Coast ranges, in parts of the Okanagan Range and northward to Wells Gray Park.

5. *Salix stolonifera* Coville

Prostrate alpine shrub, with its stems sometimes buried. **Leaves** broadly ovate, rounded at apex and base, or broadly tapered to base, 15-25 mm long, entire; hairy when young, becoming glabrous or ciliate at maturity; glaucous beneath. **Catkins** coetaneous, on leafy lateral branchlets, up to 3 cm long. **Bracts** 1.5 mm long, rounded to truncate, dark brown to black, long-hairy. **Capsules** sessile, normally glabrous, 5-7 mm long, with styles 1-2 mm long. Plate 12, figure 1.

Range: Northern British Columbia and southern Yukon to the Aleutian Islands, mainly in the mountains along the coast. In British Columbia, in the Coast Range and inland to the Liard Plateau, in alpine tundra.

6. *Salix phlebophylla* Andersson

Dwarf cushion-forming alpine shrub, with partly subterranean, rooting, smooth stems. **Leaves** ovate to elliptic, 7-20 mm long, tapered to base, obtuse to rounded at apex, green and shiny, with veins raised in both surfaces; neither deciduous nor evergreen, the old dead leaves persisting in skeletonized form, covering the stem. **Flowers** in catkins up to 2.5 cm long. Bracts up to 1.3 mm long, purple to blackish, thinly long-hairy; nectar gland one, 0.4-1.0 mm long; stamens 2, glabrous. **Capsules** 3-5 mm long, purplish, thinly hairy with short, wavy hairs, on a stipe 0.4-1.0 mm long; style 0.3-1.0 mm long, stigmas divided, 0.3 mm long. Plate 12, figure 2.

Range: Arctic-alpine, Alaska to Northwest Territories. In Yukon, found within 8 km of the British Columbia border in the Cassiar Mountains. Not yet found in British Columbia, but it may be expected.

7. *Salix rotundifolia* Trautvetter

Dwarf prostrate shrublet. Stems slender, shiny brown, often buried. **Leaves** crowded, circular to elliptic, smooth, entire, shiny, the veins often raised above; ciliate-margined but otherwise glabrous; green beneath, 4-15 mm long, cuneate to rounded at base, rounded to emarginate at apex, sometimes persistent but not skeletonized. **Catkins** coetaneous or serotinous, on short branchlets with 2 leaves, up to 2 cm long, with up to 15 flowers. Bracts 1.5-3 mm long, broadly obovate, dark purplish to brown, paler and glabrous dorsally; with long hairs ventrally, projecting beyond the margin. Gland 1 per flower, up to 0.8 mm long. Stamens 2, glabrous. **Capsules** 4-7 mm long, glabrous or with a few apical hairs, reddish brown to purplish, on a stipe 0.4-0.8 mm long; the style 0.5-1 mm long; stigmas divided. Plate 12, figure 3.

Range: Alaska and Yukon to British Columbia border, but not yet found in British Columbia. Also in the Rocky Mountain states. It may be looked for in alpine situations in northern British Columbia.

8. *Salix exigua* Nuttall   Sandbar Willow, Coyote Willow
Silverleaf Willow, Dusky Willow

Variable shrub, often of greyish appearance; with clustered or scattered, slender stems arising from widely spreading, shallow root systems. **Leaves** linear, acute, 5-12 cm long, entire to distantly and shallowly toothed; commonly greyish with fine, silky hairs on the lower surfaces or on both surfaces of young leaves. **Catkins** 3-8 cm long, terminal and also sometimes lateral, on well-developed, leafy lateral shoots. Bracts ovate to lanceolate, acute to blunt, yellowish, deciduous early, with scattered long, silky hairs. Stamens 2, their filaments hairy toward their bases. **Capsules** narrowly ovoid, 3-8 mm long, hairy or glabrous. Stigmas sessile, stout, divided, about 0.5 mm long. Plate 13, figures 1-3.

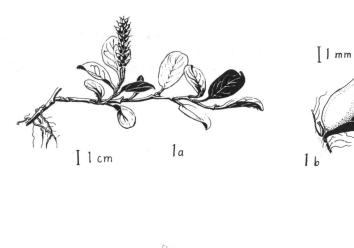

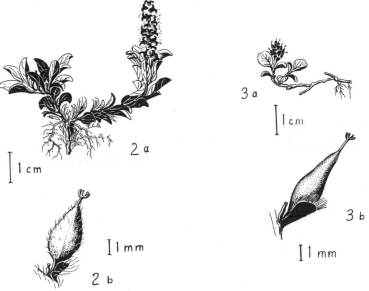

Plate 12 Prostrate Alpine Willows. Habit (a) and fruit (b). Figures:
(1) *Salix stolonifera;* (2) *S. phlebophylla;* (3) *S. rotundifolia.*

Key to the Subspecies of *S. exigua*

1a Capsules 5-8 mm long, stipitate, nearly or quite glabrous. Catkins rather loose, up to 8 cm long. Bracts variable, but tending to be pointed. Leaves toothed, rather hairy and veiny

.... **subsp.** *interior* **(Rowlee) Cronquist**

1b Capsules 3-5 mm long, more or less sessile. Catkins 3-5 cm long.

2a Leaves entire or with few scattered teeth, very slender, silky-hairy on both surfaces. Bracts narrow and pointed. Capsule hairy

.... **subsp.** *exigua*

2b Leaves serrate, relatively wider than in subsp. *exigua,* glabrescent when mature. Bracts blunt. Capsules glabrous

.... **subsp.** *melanopsis* **(Nuttall) Cronquist**

Range: Typically colonizing sand and gravel bars and banks of rivers and streams. Widespread and common from the Coast and Cascade ranges eastward across North America. Subsp. *interior* is boreal in distribution, in British Columbia mainly east of the Rocky Mountains and on the Liard River. Subsp. *melanopsis* (Dusky Willow) occurs in the southern interior and on southern Vancouver Island.

9. *Salix sessilifolia* Nuttall Sandbar Willow

Colonial shrub spreading by extensive shallow roots giving rise to erect shoots at intervals; branchlets slender, grey-hairy when young, becoming darker grey and glabrescent in age. **Leaves** greyish, lanceolate to elliptic, 4-8 cm long and 3-8 times as long as wide, greyish-hairy when young, becoming glabrescent above in age, usually entire; midvein with a minute dark callus tip; petioles less than 3 mm long; stipules minute, soon deciduous. **Catkins** appearing with or after the leaves emerge, terminating well-developed leafy lateral branchlets and sometimes also in leaf axils; staminate catkins 2-4 cm long, pistillate catkins 3-10 cm long; bracts yellow, long-haired, sometimes shed early; stamens 2, long-haired at base; **Capsules** short-pedicelled, 3-5 mm long, typically with long, stiff hairs; style short, up to 0.5 mm long, simple or divided, with deeply divided slender stigmas. Plate 13, figure 4.

Range: Southwestern British Columbia to Oregon. In British Columbia, along the lower Fraser River between New Westminster and Hope; colonizing sand and gravel bars in rivers.

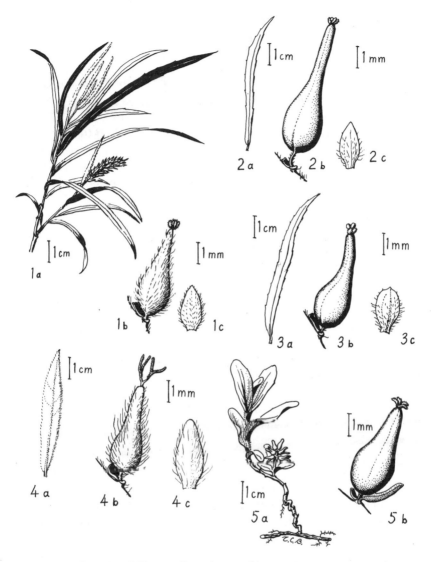

Plate 13 Sandbar Willows. Branches and leaves to a common scale; capsules and bracts to a common scale. Figures: (1) *Salix exigua* subsp. *exigua:* 1a, branch with foliage and female catkin; 1b, capsule; 1c, bract; (2) *S. exigua* subsp. *interior:* 2a, leaf; 2b, capsule; 2c, bract; (3) *S. exigua* subsp. *melanopsis:* 3a, leaf; 3b, capsule; 3c, bract; (4) *S. sessilifolia:* 4a, leaf; 4b, capsule; 4c, bract; (5) *S. setchelliana:* 5a, shoot with foliage and fruiting catkin; 5b, capsule and bract.

10. *Salix setchelliana* Ball

Low, straggling shrub up to 30 cm tall, forming colonies by means of long, shallow root systems giving rise to scattered shoots. Branchlets tomentose at least when young, greyish to purplish. **Leaves** sessile, oblong to elliptic, rounded at ends, entire or minutely toothed, fleshy or rubbery in texture, pale green above, slightly glaucous beneath, often purplish along the midvein; stipules absent or minute. **Catkins** with the leaves, on short, leafy branchlets, up to 2 cm long; bracts pale yellowish brown, glabrous or ciliate; stamens 2, glabrous. **Capsules** ovate, glabrous 5-7 mm long; pedicel about 0.5 mm long; style none or up to 0.3 mm long. Plate 13, figure 5.

Range: Yukon River basin and southern Alaska coast and on the Alsek River in extreme northwestern British Columbia; colonizing sand and gravel bars and beaches, as do the other sandbar willows, which it replaces locally.

11. *Salix amygdaloides* Andersson   Peachleaf Willow

Tree up to 15 m tall, with yellowish brown bark. Bud scales with overlapping margins (unique among our species in this respect). **Leaves** lanceolate, acuminate, serrulate, pale or glaucous beneath, 50-100 mm long; petioles up to 25 mm long. **Catkins** appearing with leaves, short-pedunculate, lax, 4-8 cm long; bracts yellow, hairy ventrally; stamens usually 5, hairy at base. **Capsules** lanceolate, glabrous, yellow, 5 mm long; style short; stigmas distinct. Plate 14, figure 1.

Range: From British Columbia to the Atlantic coast and southward to Texas. In eastern British Columbia, westward to the Okanagan Valley and the upper Fraser River and northward to the Peace River, growing along streams and lakeshores.

12. *Salix alba* L. var. *vitellina* (L.) Stokes   Golden Willow
                                                              White Willow

Small tree or big coarse shrub with ascending, bright yellow branches and twigs. **Leaves** 5-10 cm long, lanceolate, tapering at base and acuminate apex, finely serrate, with greyish silky hairs on both surfaces at first, often becoming glabrous later; glaucous beneath; petioles 5-8 mm long, not glandular; stipules soon deciduous. **Catkins** appear with the leaves, on leafy branchlets, 2.5-6 cm long; glands 1 per flower; stamens 2; ovary at first sessile, glabrous, becoming a stipitate capsule in fruit; style short. Plate 14, figure 2.

Range: This native of Europe is occasionally planted and has escaped locally, as at Victoria and Vernon.

Salix × rubens Schrank (*S. alba × fragilis*)   Golden Willow

This horticultural hybrid resembles *S. alba*, but is usually a coarse shrub with ascending greyish yellow branches; twigs and young leaves quickly becoming glabrous; ovary stipitate, the stipe becoming twice as long as the gland.

Range: This hybrid is widely planted and has been found wild in the interior of British Columbia, at places such as Vernon and Prince George and on the west coast of Vancouver Island.

13. *Salix lucida* Muhl   **Shining Willow**
(including *S. lasiandra* Bentham)   **Pacific Willow**

Big shrub or tree up to 15 m tall, with yellowish grey fissured bark on old trunks. Twigs yellowish, glabrous or pubescent. Buds smooth. **Leaves** lanceolate with attenuate apex, deep green and shiny above, serrulate on margins, with projecting glands at the junction of blade and petiole. Stipules small to large, acute, glandular-toothed. **Catkins** on short, leafy branchlets, up to 10 cm long. Bracts pale yellowish, hairy, deciduous early. Stamens usually 5, hairy at base or glabrous. Glands 2 in staminate flowers, often only 1 in pistillate flowers. **Capsules** glabrous, pale brown, 5-7 mm long, on pedicels 1.5-2 mm long, with a short style. Plate 14, figure 3.

Key to the Subspecies of *S. lucida* in British Columbia
1a   Leaf glaucous beneath ...... subsp. *lasiandra* (Bentham) E. Murray
1b   Leaf green beneath .... subsp. *caudata* (Nuttall) E. Murray

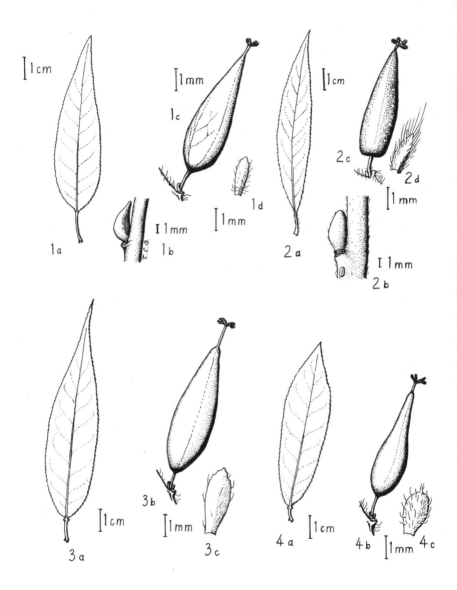

Plate 14 Willows. Leaves to a common scale; capsules and bracts to a common scale; buds to a common scale. Figures: (1) *Salix amygdaloides;* (2) *S. alba;* (3) *S. lucida;* (4) *S. serissima.* 1b, 2b, buds; 1c, 2c, 3b, 4b, capsules; 1d, 2d, 3c, 4c, bracts.

Range: Our western subspecies are found along streams and lakeshores throughout British Columbia and much of the prairie region eastward to Saskatchewan and southward to California. Subsp. *lucida* ranges from Manitoba to Labrador and southward to Delaware.

14. *Salix serissima* (Bailey) Fernald   Autumn Willow

Shrub up to 4 m tall with glabrous, shiny, yellowish brown twigs. **Leaves** 5-10 cm long, 3-4 times as long as wide, elliptic to obovate or oblanceolate, acute to obtuse at apex, acute to rounded at base, serrulate-margined, pale green or glaucous beneath, with glands at the top of the petioles. **Catkins** appearing with or after the leaves emerge, staminate catkins 1-2 cm long, pistillate catkins 3-4 cm long; bracts yellowish, short-haired, up to 2.5 mm long, soon deciduous; glands 2; stamens 5, hairy at base. **Capsules** glabrous, up to 10 mm long, on a stipe 1-2 mm long; style short. Plate 14, figure 4.

Range: Northeastern British Columbia and across the Northwest Territories to the Atlantic coast and southward to Colorado, in stream-bank thickets and other wet sites; rare in British Columbia (Dawson Creek).

15. *Salix glauca* L.   Glaucous Willow

Variable shrub 0.5-2 m tall or more, with greyish hairy twigs and pale green foliage, the bush appearing pale grey-green from a distance. **Leaves** pale green above, glaucous beneath, elliptic to lanceolate, usually entire, greyish-hairy when young, often becoming glabrescent, on petioles 4-10 mm long; stipules variable in size and persistence. **Catkins** appearing with the leaves on short, leafy branchlets, 4-5 cm long; bracts pale, yellowish brown, grey-hairy; glands 1 or 2; stamens 2; ovary densely white-hairy. **Capsules** hairy or sometimes becoming glabrous; stipe up to 1.5 mm long; style 0.5-1 mm long, sometimes divided. Plate 15, figure 1.

Range: Circumboreal; throughout British Columbia east of the Coast Range, especially abundant northward; typically in well-drained semi-open places from moderate elevations to alpine levels.

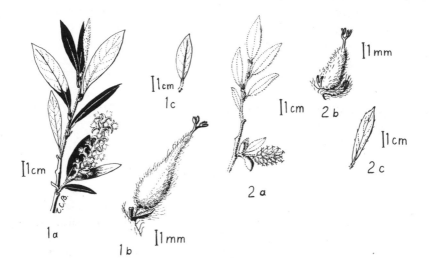

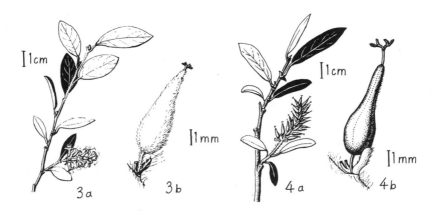

Plate 15 Willows. Branches, leaves, and catkins to a common scale; capsules with bracts to a common scale. Figures: (1) *Salix glauca* var. *acutifolia:* 1a, branch; 1b, capsule and bract; 1c, leaf of var. *villosa;*
(2) *S. brachycarpa* subsp. *brachycarpa:* 2a, branch; 2b, capsule and bract; 2c, leaf of subsp. *niphoclada;* (3) *S. athabascensis:* 3a, branch; 3b, capsule; (4) *S. raupii:* 4a, branch; 4b, capsule and bract.

Key to the Varieties of *S. glauca* in British Columbia (adapted from Argus [1])

1a Shrub 0.3-3 m tall. Leaves 2.4-5 cm long, elliptic to obovate, 1.6-3 times as long as wide, pubescent on both sides. Stipules minute or up to 6 mm long, inconspicuous. Pistillate catkins 2-4 cm long. Rocky Mountains to Hudson Bay and southward to New Mexico
. . . . **var *villosa* (Hooker) Andersson**

1b Habit various. Leaves 4-10 cm long, obovate or narrowly so, 2.8-4 times as long as wide. Stipules 4-17 mm long, prominent. Pistillate catkins 3.5-7 cm long.

    2a Shrub 0.9-4.5 m tall. Leaves relatively deep green and glabrescent above, villous to sometimes glabrescent beneath. Petioles 3-16 mm long. Stipules very prominent. Branchlets pubescent or villous. Pistillate catkins stout, long-cylindrical. Bracts tawny to light brown. Central and northern British Columbia, Yukon, Mackenzie Valley and central and northern Alaska . . . . **var. *acutifolia* (Hooker) Schneider**

    2b Shrub prostrate or up to 0.9 m high. Leaves light green above, pubescent on both sides, becoming glabrescent above, never villous beneath. Petioles 2-10 mm long. Stipules variable. Branchlets densely villous. Internodes generally short. Pistillate catkins short, narrowly cylindrical. Bracts often dark brown. Extreme northwestern British Columbia and northwestward through Alaska and westward to northern Europe . . . . **var. *glauca***

Note: These three varieties intermingle and intergrade extensively where their ranges overlap.

16. *Salix brachycarpa* Nuttall

Low, erect shrub, up to a metre tall, with grey-hairy branches; buds minute. **Leaves** 1-4 cm long, narrowly elliptic to oblanceolate, acute at apex, acute to rounded at base, entire, densely hairy beneath with long straight hairs, grey-hairy, sometimes becoming glabrescent above; petioles up to 4 mm long, often shorter than the buds. **Catkins** coetaneous, on short, leafy branchlets less than 2 cm long; bracts pale brownish, often dark at tips, hairy; glands 1 or 2 per flower; stamens 2, glabrous. **Capsules** 3-5 mm long, grey-woolly, sessile or nearly so; styles up to 0.8 mm long, stigmas shorter. Plate 15, figure 2.

Key to the Subspecies of *S. brachycarpa*

1a Mature leaves grey-hairy above, elliptic, 1-3 cm long, 1.5-3 times as long as wide; petioles shorter than mature buds; catkins nearly globular, up to 2 cm long; capsules sessile . . . . **subsp.** *brachycarpa*

1b Mature leaves glabrescent above, lanceolate to oblanceolate, 3-5 cm long, 3-4 times as long as wide; petioles longer than mature buds; catkins cylindrical, up to 5 cm long; capsules on stipes up to 0.5 mm long . . . . **subsp.** *niphoclada* **(Rydberg) Argus**

Range: Alaska to Quebec and southward in the mountains to the Rocky Mountain states. Widespread in the interior of British Columbia at moderate to alpine levels, on open rocky sites and bare alluvial flats, commonly associated with limestone.

17. *Salix athabascensis* Raup   Athabasca Willow

Erect shrub 60 cm tall, with finely grey-hairy young twigs. **Leaves** 25 mm long, elliptic or obovate, entire or minutely toothed, acute or acuminate; base obtuse or rounded; green and thinly pubescent, becoming glabrous above; glaucous with thinly appressed hairs beneath, becoming glabrous with age. **Catkins** 2 cm long, on leafy peduncles; bracts 1-1.3 mm long, obtuse or rounded, brown, silky-hairy. **Capsules** 5-6 mm long, covered with short, appressed hairs until mature, becoming glabrous, on stipes 1-1.5 mm long; style about 0.3 mm long. Plate 15, figure 3.

Range: Alaska to Hudson Bay in muskeg and swamps, southward in British Columbia at low elevations to about lat. 58°N, but uncommon.

18. *Salix raupii* Argus   Raup's Willow

Shrub 1-2 m tall, with glabrous, glossy, chestnut-brown branchlets. **Leaves** pale green as in *S. glauca,* but glabrous from the start. **Flowers** with bracts pale, becoming brownish with age, glabrous or with sparse silky hairs, and ciliate when young; ovary glabrous or almost so; stigmas 4-lobed, red. **Capsules** brown, more or less glabrous. Plate 15, figure 4.

Range: Little known; local in the northern Rocky Mountains and Cassiar Mountains (McDonald Creek on the Alaska Highway).

19. *Salix barclayi* Andersson   Barclay Willow

Variable shrub to 4 m tall, or low and spreading; branches and twigs blackish, tomentose to glabrous. **Leaves** 40-77 mm, oval to obovate, acute or sharp-pointed at apex, acute to rounded at base, serrulate to almost entire, slightly hairy above on veins when young, glaucous beneath. Stipules large and early deciduous. **Catkins** stout on leafy peduncles, 2.5-7.5 cm long; bracts greyish brown, mostly acute, thinly to densely villous. **Capsules** glabrous in var. *barclayi or* thinly tomentose in var. *latiuscula* Andersson; 5-8 mm long; stipes 0.5-1.5 mm long; styles 1-1.5 mm long. Plate 16, figure 1.

Range: Alaska to Washington along the mountain ranges, in a wide range of habitats and altitudes.

20. *Salix commutata* Bebb   Variable Willow

Shrub to 3 m tall with dark, stoutish, variably hairy twigs. **Leaves** elliptic to obovate, 50-75 mm long, abruptly sharp-pointed at the apex, entire or minutely glandular-serrate; tomentose on both sides, becoming less hairy with age; green on both surfaces. **Catkins** stout, 3-5 cm long, on leafy peduncles; bracts brown, woolly. **Capsules** 5-7 mm long, mostly glabrous; stipe 1 mm long; style 1-1.5 mm long; stigmas short; stamens 2, filaments glabrous, free. Plate 16, figure 2.

Key to the Varieties of *S. commutata*
1a   Leaves entire, densely hairy at least when young.
    2a   Capsule glabrous . . . . **var.** *commutata*
    2b   Capsule sparsely hairy . . . . **var.** *puberula* Bebb
1b   Leaves serrulate, sparsely hairy . . . . **var.** *denudata* Bebb

Range: From Alaska to the Northwest Territories and southward to California and the Rocky Mountain states in the mountain ranges. Widespread in British Columbia, on alpine tundra and on gravelly river banks and lakeshores at lower elevations.

21. *Salix pseudomonticola* Ball   Serviceberry Willow
(*S. monticola* Bebb, in part;
S. padophylla Rydberg)

Shrub to 4 m tall; branchlets yellowish to red or brown, shiny, usually glabrous. **Leaves** obovate to narrowly or broadly ovate, 40-50 mm long by 25 mm broad, mostly cordate at base, acute or short acuminate at apex, coarsely glandular, bluntly serrate, glaucous beneath, glabrous and strongly veined on both surfaces, densely pubescent in some southern forms. **Catkins** sessile or with very short stalk, with or without leafy bracts; pistillate 3-8 cm long; bracts broadly oblanceolate, obtuse, brown with long hairs. **Capsules** 6-8 mm long, glabrous, yellowish; style under 1 mm long; stamens 2, filaments glabrous, free. Plate 16, figure 3.

Range: Alaska to Northwest Territories and southward to the Rocky Mountain states. Widely scattered in the interior of British Columbia, in moist sites at low to moderate elevations in the forested regions.

22. *Salix prolixa* Andersson   Mackenzie Willow
(*S. rigida* Muhlenberg;   Yellow Willow
S. cordata Muhlenberg; *S. lutea* Nuttall;
S. mackenzieana Barratt in Andersson;
S. watsonii Bebb; *S. macrogemma* Ball)

Highly variable shrubs or small trees up to 6 m tall; twigs elongate, reddish brown to yellowish, pubescent when young, usually becoming glabrous. **Leaves** ovate to lanceolate, rounded to cordate at base, acuminate at apex, glandular-toothed, glabrous except for a band of short hairs along the upper side of the petiole and midvein, glaucous beneath; stipules usually prominent, kidney-shaped. **Catkins** on branchlets 4-8 mm long bearing 2-3 small leaves, lax, 25-70 mm long; the catkin axis often densely woolly; bracts about 1 mm long, brown, usually long-hairy. **Capsules** 4-7 mm long, glabrous, on a stipe 2.5-4 mm long; style 0.2-0.7 mm long; stamens 2, glabrous. Plate 16, figure 4.

Range: Transcontinental: Yukon to Newfoundland and southward to Arizona. Throughout British Columbia, by lakeshores and in other moist sites.

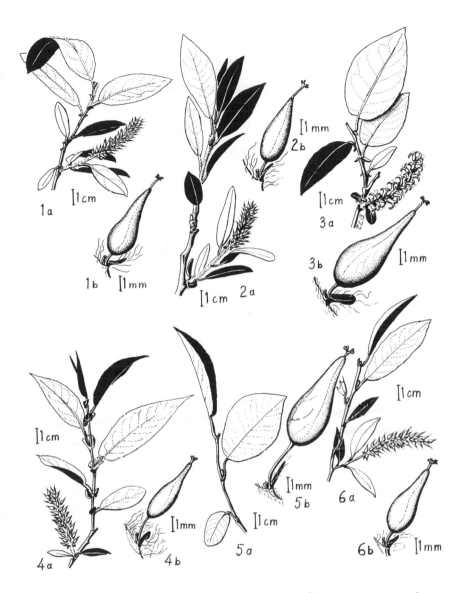

Plate 16 Willows. Branches, foliage, and catkins (a) to a common scale; capsules and bracts (b) to a common scale. Figures: (1) *Salix barclayi;* (2) *S. commutata;* (3) *S. pseudomonticola;* (4) *S. prolixa;* (5) *S. pyrifolia;* (6) *S. hastata.*

23. *Salix pyrifolia* Andersson   Balsam Willow

Upright shrub up to 3 m tall, with shiny reddish brown to greenish branches. **Leaves** 20-60 mm long, obovate to ovate or lanceolate, acute to obtuse at apex, cordate at base; thin, sparsely hairy when young; green above, glaucous beneath at maturity, finely glandular-serrate, on long slender petioles. **Catkins** coetaneous, on short, small-leafed branchlets in var. *laeta* Andersson, almost sessile in var. *obscura* Andersson; bracts lanceolate, up to 2 mm long, brown, thinly hairy. **Capsules** glabrous, 4-9 mm long, on stipes that are up to 4 mm long and much longer than the minute gland; styles 0.4-1 mm long. Plant often with a balsamic fragrance. Plate 16, figure 5.

Range: Boreal Forest Region from the Yukon to Newfoundland. In British Columbia, rather scarce; southward east of the Rocky Mountains to the Peace River and in the Prince George area; found in marshy stream and lake margins and in fens.

24. *Salix hastata* L. var. *farriae* (Ball) Hulten   Farr Willow
(*S. farriae* Ball)

Small shrub with glabrous, bright-red branches. **Leaves** variable, ellipticoblanceolate, acute at apex, entire or nearly so, to 50 mm long, glabrous, slightly glaucous beneath; petioles slender. **Catkins** short-pedunculate, 2.5-5 cm long; bracts oblong, obtuse and greyish-brown at apex, glabrous outside. **Capsules** lanceolate, glabrous, pedicel 1-1.5 mm long; style 0.3-0.5 mm long, stigmas short. Plate 16, figure 6.

Range: *S. hastata* ranges from Hudson Bay to Alaska and across northern Eurasia. Shrubs in extreme northwestern British Columbia may be var. *hastata*. Var. *farriae* is uncommon in central and southeastern British Columbia to the Rocky Mountain states. The species is found in a variety of open habitats, usually at subalpine elevations.

25. *Salix bebbiana* Sargent   Bebb Willow

Shrub with several very branchy stems, or small tree to 6 m tall, with twigs hairy to smooth. **Leaves** narrowly elliptic and acute at both ends to broadly oblanceolate and short-acuminate, 4-8 cm long, usually entire, more or less pubescent or almost glabrous in age, glaucous beneath. **Catkins** precocious to coetaneous, on small-leafed, short branchlets, open in form, to 6 cm long, lax; bracts yellowish or brown, acute, 1-2 mm long; stigmas nearly sessile, deeply divided. **Capsules** 6-10 mm long, thinly pubescent; pedicels 2-5 mm, pubescent, yellow. Plate 17, figure 1.

Key to the Varieties of *S. bebbiana*
1a Capsule and bract hairy.
   2a Lower leaf surface puberulent with upstanding hairs. Veins raised
     and conspicuous . . . . **var.** *bebbiana*
   2b Lower leaf surface thinly appressed-hairy to glabrescent; veins
     inconspicuous . . . . **var.** *perrostrata*
1b Capsule glabrous; bracts glabrous or sparsely ciliate . . . . **var.** *depilis*

Range: Transcontinental; from Alaska to Newfoundland, mainly in the Boreal Forest Region, but extending southward in the west to Washington and Montana. The commonest upland willow at moderate altitudes everywhere east of the Coast range in British Columbia. Var. *depilis* Raup occurs from the latitude of Prince George northward into the Mackenzie Valley. Var. *perrostrata* (Rydberg) Schneider is probably our commonest variety, but appears not to have a distinct geographic range in this province.

26. *Salix discolor* Muhlenberg   Pussy Willow

Coarse shrub or small tree; twigs glabrous or soon becoming so. **Leaves** elliptic to broadly oblanceolate, 3-10 cm long by 1-3 cm wide, acute at both ends, crenate to entire, glaucous beneath, glabrous or brownish-hairy when young, becoming glabrous in age. Petioles 0.5-3 cm long. Stipules small, or prominent and ovate on vigorous shoots. **Catkins** precocious, sessile, silky-hairy when young; staminate catkins up to 5 cm long; pistillate catkins elongating up to 12 cm long by 3 cm thick in fruit. Bracts dark reddish brown to black except for a pale base, 2-3 mm long, with straight hairs equally long. Glands 1, 0.5 mm long, truncate. Stamens 2, with short hairs toward the base. **Capsules** 5-12 mm long, with short, wavy hairs, on a stipe 1.5-2.4 mm long, the style about 0.8 mm long with short hairs at least at the base. Plate 17, figure 2.

Range: An upland willow of wooded and open areas, mainly of the Boreal Forest Region; ranging from the interior of British Columbia eastward across the continent. Stations at its western limit include Swan Lake in the Cassiar Mountains, Smithers and Kelowna.

27. *Salix scouleriana* Barratt *ex* Hooker   Scouler's Willow
(*S. nuttallii* Sargent)

Big shrub or small tree 6-9 m tall, with smoothish grey bark and greenish to reddish brown, pubescent or glabrous twigs. **Leaves** 40-120 mm long, obovate or oblanceolate, tapered to base; obtuse or rounded or sometimes abruptly acute, at apex; entire or slightly crenate, rather thick and firm in texture, glabrous above, glaucous and commonly pubescent beneath, with silvery or brown hairs, short-petioled. **Catkins** 2-6 cm long, precocious and sessile in var. *scouleriana*, and coetaneous on short leafy shoots in var. *coetanea* Ball; bracts obovate, black, 4-5 mm long, long-hairy. **Capsules** 7-9 mm long, short-hairy, with stipes up to 1 mm long; styles up to 0.5 mm long. Plate 17, figure 3.

Range: Alaska to Manitoba and southward to California and Arizona; throughout British Columbia. Var. *scouleriana* is the common upland willow on the coast, where it often forms a small crooked tree. The species also occurs, mainly as var. *coetanea,* across the interior, in dry wooded habitats at moderate elevations.

28. *Salix hookeriana* Barratt *ex* Hooker   Hooker's Willow

Tall shrub or small tree up to 6 m tall, with rough dark grey bark; twigs coarse, densely pubescent, brittle-based. **Leaves** broadly elliptic to obo vate, obtuse at apex, the margin crenate to entire, dark greyish green and tomentose or glabrescent above, densely tomentose beneath, 4-12.5 cm long. **Catkins** precocious, sessile or on short peduncles, 4-7.5 cm long by up to 2.5 cm thick, densely long-hairy; bracts dark, up to 5 mm long, copiously long-hairy. **Capsules** 6-9 mm long, glabrous in var. *hookeriana,* and more commonly here, tomentose in var. *tomentosa* Henry; capsules in both varieties on pedicels 1-3 mm long, with styles 1-2 mm long; the stigmas undivided, short or long. Plate 17, figure 4. Photo 19.

Range: Alaska to California along the coast. In British Columbia, close to the coast, in marshes and along swampy lakeshores.

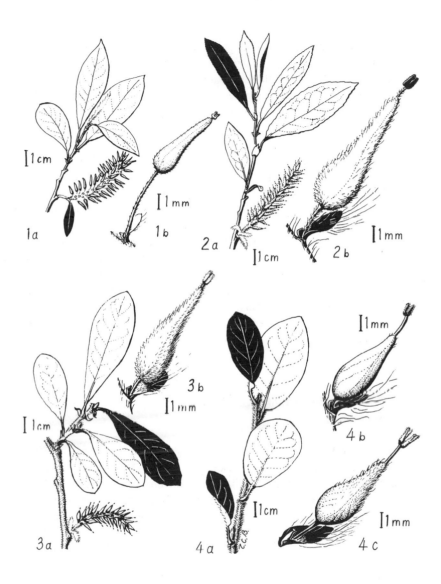

Plate 17 Willows. Branches, leaves, and catkins (a) to a common scale; capsules and bracts (b and c) to a common scale. Figures: (1) *Salix bebbiana;* (2) *S. discolor;* (3) *S. scouleriana;* (4) *S. hookeriana:* 4b, capsule of var. *hookeriana;* 4c, capsule of var. *tomentosa.*

29. *Salix alaxensis* (Andersson) Coville   Alaska Willow

Tall shrub or tree to 8 m tall, or a low alpine shrub, with coarse branches and twigs. **Leaves** elliptic to oblanceolate, 30-90 mm long by 20-30 mm wide, the apex acute or obtuse, the base acute, surface glabrous or nearly so above, densely white-felted beneath; with conspicuous lanceolate stipules. **Catkins** precocious, erect, sessile, 3-15 cm long; bracts 3 mm long, black with white hairs. **Capsules** tomentose, 6-8 mm long, sessile or nearly so; styles 2 mm long. Plate 18, figure 1.

Key to the Varieties of *S. alaxensis*
1a Twigs permanently densely white-woolly . . . . **var.** *alaxensis*
1b Twigs glabrous or almost so, with a bluish waxy bloom
    . . . . **var.** *longistylis*

Range: Var. *alaxensis* from Hudson Bay to central Siberia, northward to the Arctic coast and southward in the Rocky Mountains to about lat. 53°N. Var. *longistylis* (Rydberg) Schneider from Alaska to central British Columbia. In British Columbia, var. *alaxensis* is the commoner variety at alpine and high subalpine levels, while var. *longistylis* is commoner at low to subalpine levels. Both varieties show a preference for open and semi-open situations, often on rocky soils.

30. *Salix candida* Fluegge *ex* Willdenow   Hoary Willow
                                                       Sage-leafed Willow

Upright shrub up to a metre tall; twigs white-pubescent, becoming glabrescent and reddish with a bloom. **Leaves** 25-50 mm long, oblong or lanceolate, acute at apex, thinly hairy and becoming glabrescent; dark dull greyish green above, densely white-woolly beneath, the margin revolute, entire. **Catkins** precocious, 5-10 cm long, sessile, densely flowered; bracts 1 mm long, brown, woolly. **Capsules** 6-7 mm long, white-woolly, on stipes 1 mm long; styles 1 mm long, they and the stigmas red. Plate 18, figure 2.

Range: Alaska to Labrador, southward in the mountains to Colorado. Widespread in the interior of British Columbia in wet and seeping areas, particularly where the water is calcareous or saline.

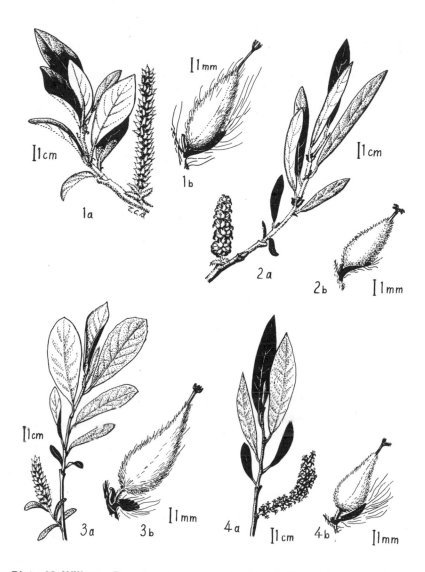

Plate 18 Willows. Branches, leaves, and catkins (a) to a common scale; capsules and bracts (b) to a common scale. Figures: (Top) Leaves white-felted beneath: (1) *Salix alaxensis;* (2) *S. candida.* (Bottom) Leaves satiny beneath: (3) *S. sitchensis;* (4) *S. drummondiana.*

31. *Salix sitchensis* Sanson *ex* Bongard   Sitka Willow

Shrub or small tree with thin greyish-brown bark; young twigs and branches usually densely hairy, becoming glabrous with age. **Leaves** 40-100 mm long, oblong or oblanceolate, acute or rounded at apex, tapering toward base, entire or obscurely glandular-toothed; slightly pubescent to glabrous above, with appressed shiny hairs beneath giving the surface a satinlike lustre; petioles short, hairy; stipules large or minute, deciduous early. **Catkins** 2.5-7.5 cm long, slender, ascending, precocious or coetaneous on short branchlets with reduced leaves; bracts brown, long-hairy; staminate flower typically with one stamen. **Capsules** 4-6 mm long, densely hairy with shiny hairs; stipe 0.5-1.4 mm long; style 0.3-1.2 mm long. Plate 18, figure 3.

Range: Alaska to California along the coast and inland in British Columbia. Throughout British Columbia, but commonest on the coast, mainly along shores and streams.

32. *Salix drummondiana* Barratt   Drummond's Willow
(*S. subcoerulea* Piper)

A shrub 1-2.5 m tall, twigs and branches with a bluish bloom. **Leaves** 25-50(-75) mm long, oblong-lanceolate, usually oblanceolate, to acute or obtuse at apex and base, entire or nearly so, green and thinly pubescent above, silvery beneath with a dense, short, silky pubescence. **Catkins** precocious, the pistillate ones up to 8 cm long, densely flowered; bracts 1-2.5 mm long, dark, acute-tipped, long-hairy; glands 0.4-0.8 mm long, 1 per staminate flower, 2 per pistillate flower. **Capsules** 4-5 mm long, silvery-silky pubescent; stipe up to 1.4 mm long; style to 1.3 mm long. Plate 18, figure 4.

Range: Yukon to Alberta, Montana and Oregon. In British Columbia, from the Coast and Cascade ranges eastward; found in willow thickets by swampy lake or stream shores, usually at subalpine levels.

33. *Salix geyeriana* Andersson   Geyer's Willow

Shrub up to 3 m tall; twigs glabrous, blackish with a bluish bloom in var. *geyeriana*, and brown (darkening on drying) and mostly nonbloomy in var. *meleina* Henry. **Leaves** lanceolate, 3-8 cm long, acute at ends, entire; green but slightly silky above, glaucous and thinly silky-pubescent beneath, often with brown hairs. **Catkins** coetaneous, 1-2.5 cm long, loosely flowered; bracts linear-oblong, obtuse, thinly pubescent, dark reddish brown. **Capsules** 3-6 mm long, thinly short-puberulent, on a stipe 1-2.5 mm long; style short or none; stigmas short. Plate 19, figure 1.

Range: Southern British Columbia to California and Colorado. In British Columbia, scarce, by lake shores: var. *geyeriana* recorded from Moyie and Logan lakes in the interior; var. *meleina* on southern Vancouver Island from Duncan southward.

34. *Salix lemmonii* Bebb   Lemmon's Willow

Very similar to *S. geyeriana*. Twigs may be pruinose or not. **Leaves** sericeous when young, becoming glabrous on both sides at maturity, on petioles 5-9(-12) mm long, with stipules often expanded, at least at later leaves. **Catkins** coetaneous; staminate catkins 1-3.5 cm long; pistillate catkins 2-6 cm long. Bracts dark brown to black, obovate to elliptic, rounded to truncate at apex, 1-1.4 mm long, less than twice as long as wide. **Capsules** (4-)5-9 mm long. Style 0.4-0.9 mm long. Plate 19, figure 2.

Range: Oregon and California to Idaho, Wyoming and Colorado, along streams and in wet meadows. This species has recently been reported by Dr G.W. Argus (personal communication, 1990) in the Saanich Peninsula of Vancouver Island, growing in mixture with *S. geyeriana* var. *meleina* and some intermediates. (I am indebted to Dr Argus for helping to identify components of this mixture.)

35. *Salix arbusculoides* Andersson

Shrub or small tree with reddish, shiny twigs, glabrous or nearly so. **Leaves** 2-6 cm long, narrowly lanceolate, acute at apex and base, serrulate or entire, glabrous above, thinly and shortly silky hairy or glabrous beneath. **Catkins** coetaneous or earlier, 5 cm long; sessile or with 2 or 3 small leaves on a short branchlet; bracts 1 mm long, blackish at least at apex, hairy. **Capsules** 4-7 mm long, clothed with short, stiff hairs, on stipes 0.5-1.2 mm long; styles up to 0.8 mm long. Plate 19, figure 3.

Range: Alaska to Hudson Bay through the Boreal Forest Region. Across northern British Columbia at low elevations, southward to the upper Fraser River and the Chilcotin Plateau.

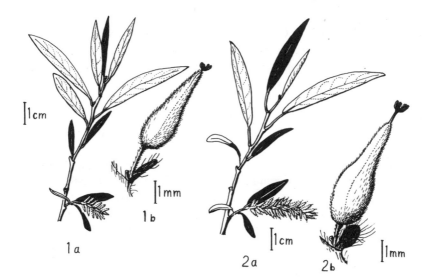

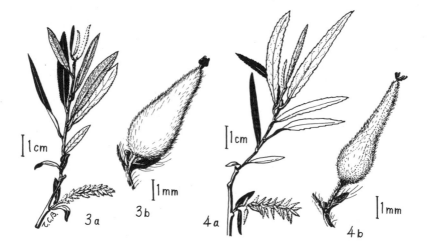

Plate 19 Willows. Branches, leaves, and catkins (a) to a common scale; capsules and bracts (b) to a common scale. Figures: (1) *Salix geyeriana;* (2) *S. lemmonii;* (3) *S. arbusculoides;* (4) *S. petiolaris.*

36. *Salix petiolaris* J.E. Smith (*S. gracilis* Andersson)

Shrub 1-4 m tall, with twigs glabrous or soon becoming so, yellowish to reddish brown, later blackening. **Leaves** linear to lanceolate, 3-7 cm long, up to 10 times as long as wide, tapering to ends, entire or serrate, glaucous beneath, silvery-silky hairy when young, becoming glabrous. **Catkins** precocious or coetaneous, sessile or on short branchlets; staminate catkins 1-2 cm long, pistillate catkins 2-4 cm long, diverging or semi-pendulous; bracts narrow, brown or yellowish, pubescent; stamens 2, hairy near base. **Capsules** silky-hairy to almost glabrous; stipe 1.5-2.5 mm long; style very short or none. Plate 19, figure 4.

Range: Northern British Columbia to Hudson Bay. In British Columbia, uncommon, from the Peace River basin northward, in streamside thickets.

37. *Salix maccalliana* Rowlee   Maccall's Willow

Erect shrub, up to 2 m tall, with reddish brown branchlets that are hairy when young, becoming glabrous. **Leaves** firm, narrowly elliptic to oblanceolate, up to 8 cm long, acute at base and apex, pubescent when young, becoming glabrous, bright green above; paler green, not glaucous beneath, the margin finely glandular-crenate; midvein and petiole yellowish. **Catkins** coetaneous, 2-5 cm long, on short, leafy branchlets; bracts 3-5 mm long, pale yellowish brown; glands 1 or 2 per flower. **Capsules** up to 12 mm long, densely white-woolly; stipe 2 mm long; styles about 1 mm long. Plate 20, figure 1.

Range: British Columbia to Quebec. Widely scattered but uncommon in the interior of British Columbia, in moist, open, low-lying sites.

38. *Salix planifolia* Pursh

Erect shrub up to 4 m tall with dark brown or purplish twigs that are sparsely hairy to glabrous. **Leaves** 2.5-7 cm long, rhombic-lanceolate to elliptic, acute at base and apex, thinly tomentose to glabrous, glaucous beneath, entire or serrulate; stipules ovate, up to 3 mm long and deciduous in subsp. *planifolia,* and lanceolate, 3-15 mm long and persistent in subsp. *pulchra* (Chamisso) Argus. **Catkins** precocious, sessile, 3-7 cm long; bracts about 2 mm long, acute at apex, blackish toward apex, hairy.

Capsules sessile or nearly so, 5-8 mm long, pubescent; styles 1-2 mm long. Plate 20, figure 2.

Range: Subsp. *planifolia* from Yukon to the Atlantic coast; widespread across the interior of British Columbia on water margins; the commoner subspecies at low elevations. Subsp. *pulchra* ranges from the Northwest Territories westward across Siberia; and southward to about lat. 56°N in British Columbia, commonly on alpine tundra, where rather low and sprawling.

39. *Salix pedicellaris* Pursh   Bog Willow

Shrub up to 1 m tall, mostly glabrous. **Leaves** oblong to oblanceolate, less than 5 cm long, mostly obtuse at ends, dull green above, usually glaucous beneath; the margins entire and sometimes revolute; the petiole short and without stipules. **Catkins** coetaneous, less than 2.5 cm long, on short, leafy branchlets; bracts about 1 mm long, yellow, nearly glabrous. **Capsules** narrowly conical, 5-10 mm long, glabrous, on stipes 2-4 mm long; styles short or none. Plate 20, figure 3.

Key to the Varieties of *S. pedicellaris*
1a   Leaves oblong, rounded at ends; capsules blunt at tip.
     2a   Leaves green beneath .... **var.** *pedicellaris*
     2b   Leaves glaucous beneath .... **var.** *hypoglauca* **Fernald**
1b   Leaves oblanceolate, acute at ends; capsule pointed at tip
     .... **var.** *tenuescens* **Fernald**

Range: Yukon to the Atlantic coast and southward into the United States; widespread in the interior of British Columbia, in bogs. The common variety in this province is var. *hypoglauca* (51).

40. *Salix myrtillifolia* Andersson

Low shrub 0.1-1.5 m tall with greyish brown branches; twigs green, with minute curled puberulence. **Leaves** 2-5 cm long, elliptic to oblong-oblanceolate, glabrous or with minute curled hairs along the upper side of the midvein and petiole; green on both sides, usually rounded at base, blunt or acute at apex, the margin serrulate; stipules none or up to 1 mm long. **Catkins** coetaneous, 2-3 cm long, on short, small-leafed branchlets;

bracts less than 1 mm long, brown, black at the rounded tip, grey-hairy. Capsules 3-7 mm long, glabrous, on a stipe 1-3 mm long; style 0.3-0.8 mm long; stigmas short, almost entire. Plate 20, figures 4 and 5.

Key to the Varieties of *S. myrtillifolia*

1a Leaves elliptic, one-half to two-thirds as wide as long, rounded at ends, 1-3 cm long, glabrous. Stipules less than 2 mm long. Stipe up to twice as long as gland. Style 0.3-0.5 mm long. Shrub usually less than 1 m tall . . . . **var.** *myrtillifolia*

1b Leaves up to 5 cm long, at least partly pubescent. Stipules 1-5 mm long. Stipe up to 3 times as long as gland. Style 0.5-0.9 mm long. Shrub commonly 1-3 m tall . . . . **var.** *cordata*

Range: Interior of British Columbia to Yukon. Var. *myrtillifolia* is primarily northern, extending southward to the Peace River basin and southward along the Coast range to southern British Columbia. It typically occurs in muskeg and Black Spruce bog-forest.

Var. *cordata* (Andersson) Dorn (var. *pseudomyrsinites* [Andersson] Ball) is the prevalent variety in the Cariboo, although it extends from northern British Columbia southeastward along the Rocky Mountains to southeastern British Columbia. It typically occurs in moist but better-drained soils than var. *myrtillifolia,* near lakes and streams. It intergrades with var. *myrtillifolia* (15).

41. *Salix boothii* Dorn   Booth's Willow

Shrub up to 6 m tall. Twigs greenish to brown, usually finely puberulent with curved hairs. **Leaves** lanceolate, rounded at base, short-acuminate at apex, 2-8 cm long by 1-2.5 cm wide, serrulate to entire, green on both surfaces, pubescent when young, becoming glabrous. **Catkins** coetaneous or barely precocious, 1-5 cm long, on branchlets up to 1 cm long, often with one or two small basal bract-leaves; the catkin axis villous with curly hairs. Bracts brown to black, 0.7-1.3 mm long, villous with curly hairs. **Capsules** glabrous, 4-6 mm long, on stipes 0.5-2 mm long, with styles 0.3-0.5 mm long. Plate 20, figure 6.

Range: Along streams and lakeshores and in meadows in the southern interior of British Columbia (Aspen Grove, Williams Lake, Fernie), and from southern Alberta to Colorado and California (7, 15).

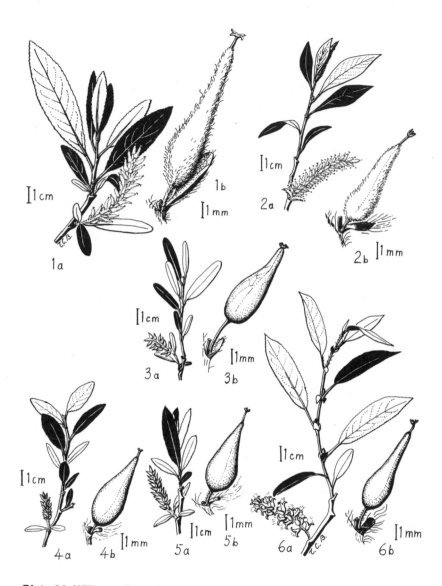

Plate 20 Willows. Branches, leaves, and catkins (a) to a common scale; capsules and bracts (b) to a common scale. Figures: (1) *Salix maccalliana;* (2) *S. planifolia;* (3) *S. pedicellaris;* (4) *S. myrtillifolia* var. *myrtillifolia;* (5) *S. myrtillifolia* var. *cordata;* (6) *S. boothii.*

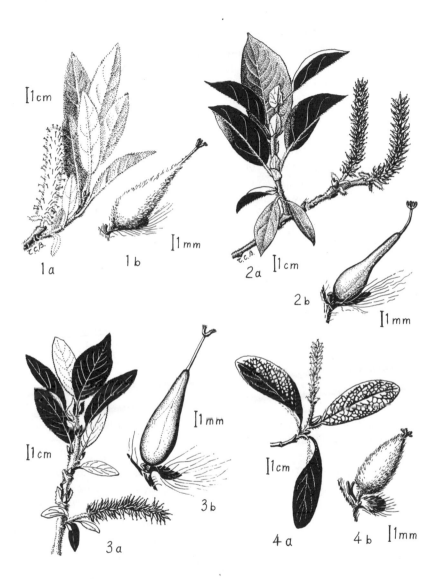

Plate 21 Willows. Branches, leaves, and catkins (a) to a common scale; capsules and bracts (b) to a common scale. Figures: (1) *Salix barrattiana;* (2) *S. tweedyi;* (3) *S. lanata;* (4) *S. vestita* var. *erecta.*

42. *Salix barrattiana* Hooker   Barratt's Willow

Erect aromatic shrub up to 1.5 m tall, with stout, dark, glandular, pubescent twigs. **Leaves** 4-9 cm long, elliptic to obovate, greyish green, persistently long-tomentose on both sides, acute at apex, acute to rounded or cordate at base. **Catkins** precocious or coetaneous, sessile, often on the ends of branches, 2.5-9 cm long, usually erect, notably white-tomentose; bracts up to 3 mm long, black, densely long-hairy. **Capsules** 5-6 mm long, white-hairy, on stipes up to 2 mm long; styles 1.2-2.5 mm long. Plate 21, figure 1.

Key to the Varieties of *S. barrattiana*
1a Stipules persistent into second year . . . . var. *marcescens* Raup
1b Stipules deciduous in first season.
    2a Leaves narrowly oblanceolate, tapered to base
    . . . . var. *angustifolia* Andersson
    2b Leaves elliptic to obovate.
        3a Leaves rounded or cordate at base, white-hairy above
        . . . . var. *barrattiana*
        3b Leaves cordate at base, glabrous and green above
        . . . . var. *latifolia* Andersson

Range: Alaska to Montana along the mountain ranges; widespread in the interior of British Columbia, at alpine and subalpine elevations east of the Coast range, often in damp alpine meadows and on limestone.

43. *Salix tweedyi* (Bebb *ex* Rose) Ball   Tweedy's Willow
(*S. barrattiana* Hooker var. *tweedyi* Bebb)

Shrub up to 2 m tall. Twigs stout, with thick, persistent hairiness. **Leaves** broadly elliptic, up to 9 cm long by 5 cm wide, rounded at base and acute at tip; the margin serrulate; stiff, thinly hairy when young, becoming glabrous except for the hairy upper surfaces of the principal veins; pale green beneath. Petiole very hairy. Stipules conspicuous, obliquely ovate, palmately veined, sometimes overlapping along the twig. **Catkins** at and near the tips of the twigs, as in *S. barrattiana,* sessile or nearly so, precocious or coetaneous. Staminate catkins about 3 cm long; pistillate catkins up to 10 cm long. Bracts 2-3 mm long, dark brown to black, with long hairs. Stamens 2, glabrous. **Capsules** 5-7 mm long, glabrous or hair-tipped, pale green. Style 1.5-2 mm long, with a 4-branched stigma about 0.5 mm long. Plate 21, figure 2.

Range: British Columbia to Montana and Wyoming; in British Columbia, found so far only near Falkland. Inhabiting stream banks and swampy lakeshores at subalpine levels (7).

44. *Salix lanata* L. subsp. *richardsonii* (Hooker) Skvortsov

Shrub to 2 m tall, broad and spreading; twigs with grey tomentum over a dark surface. **Leaves** elliptic, acute to acuminate at apex, 2.5-8 cm long and ¼ to ¾ as wide; pale green above, glaucous beneath; hairy at least when young; generally glabrous beneath at maturity, with persistent short hairs on the midvein above; stipules ovate-lanceolate, 6-25 mm long, persistent. **Catkins** before or with the leaves, almost sessile, with reduced basal leaves, the pistillate ones up to 8 cm long in fruit; bracts blackish, long-hairy; stamens 2, glabrous. **Capsules** glabrous, 5-7 mm long, nearly sessile or on a pedicel up to 1 mm long; style 1.5-3 mm long; stigmas 0.5-0.8 mm long, undivided. Plate 21, figure 3.

Range: Eastern Siberia to Hudson Bay, including British Columbia north of about lat. 56°N; along lakeshores and streams.

45. *Salix vestita* Pursh var. *erecta* Andersson   Rock Willow

Shrub up to a metre tall, with glabrous or pubescent twigs. **Leaves** 25-50 mm long, elliptic; obtuse or rounded at both ends; on stout, 2-8 mm long petioles without stipules; deep green and strongly reticulately wrinkled by impressed veins above and glaucous beneath, with silky appressed hairs at least on the raised veins. **Catkins** serotinous, terminal, 2-5 cm long, slender; bracts up to 1.5 mm long, brown, darker toward tips, hairy; nectar glands normally 2, one of them very reduced. Stamens 2. **Capsules** 3-5 mm long, ovoid, not beaked, greyish-hairy; stipe up to 1 mm long; style almost none. Plate 21, figure 4.

Range: Central British Columbia and Alberta to Montana and Oregon, in the Rocky Mountains and, less commonly, in the more western ranges, often on limestone at alpine and subalpine elevations. In British Columbia, reaching the Omineca Mountains at about lat. 55°N.

Myricaceae   Sweet Gale Family

Shrubs or small trees. Leaves alternate, simple, lobed or toothed, with or without stipules, with minute resinous dots. Buds small and scaly. Flowers unisexual, in short catkins borne on old wood or in the axils of the older leaves; no perianth; staminate flowers with 2-12 stamens, with or without basal bractlets; pistillate flowers with bractlets and unilocular ovaries with paired filiform stigmas. Fruit a nut or drupe. Our only genus is *Myrica.*

Myrica   Wax-myrtles

Leaves shallowly few-toothed, not lobed; without stipules, with minute resinous dots. Pistillate flowers with 2-6 minute bractlets shorter than the fruit. Fruit a nut or drupe.

Key to the Species of *Myrica*
1a   Flowers dioecious; leaves deciduous; fruit spotted with resin, ovoid, with prominent bracts . . . . *M. gale*
1b   Flowers monoecious; leaves evergreen; fruit covered with a whitish wax, globular, with minute bracts . . . . *M. californica*

Myrica gale L.   Sweet Gale

Dioecious, aromatic shrub 1-1.5 m tall, with dark brown glabrous branches. **Leaves** appear after the flowers; serrate toward the obtuse apex, cuneate toward an almost sessile base; 25-65 mm long, glabrous above, firm. **Catkins** conelike, on an old leafless twig. **Fruit** a nutlet scarcely equalling the 2 winglike attached bracts. Plate 22, figure 1.

Range: Circumboreal, by the margins of lakes and swamps. In British Columbia, throughout the coastal region, by subalpine lakes in the central interior and from the upper Peace River northward.

Myrica californica   California Wax-myrtle
Chamisso & Schlechtendal

A shrub or, in favourable situations, a small tree to 8 m tall. **Leaves** 5-10 cm long, oblanceolate, short-petioled, tapering to base, acute at apex, distantly serrate, glabrous, thick and evergreen. **Catkins** on older wood or in the axils of older leaves, 1-2 cm long, often with staminate flowers at bases and pistillate flowers above. **Fruit** 4-8 mm across, brownish-purple, nutlike; covered with waxy, wartlike glands, which give it a grey to whitish appearance at maturity; with 2 minute bracts at the base of the fruit. Plate 22, figure 2. Photo 20.

Range: From the west coast of Vancouver Island to California along the coast. In British Columbia, known from Ucluelet and Tofino, along the ocean shore.

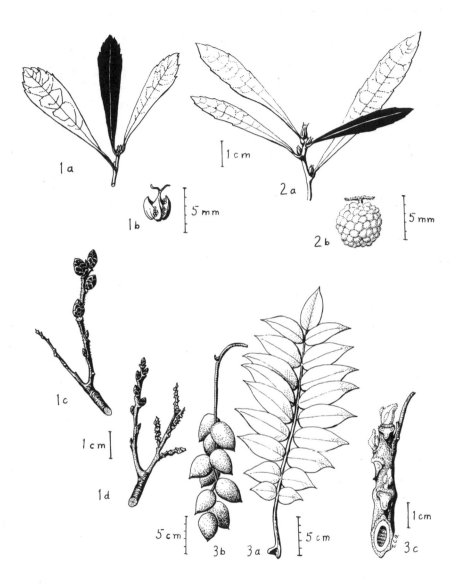

Plate 22 Wax-myrtles and Japanese Walnut. Figures: (1) *Myrica gale:* a, foliage; b, fruit; c, staminate winter twig; d, pistillate winter twig; (2) *M. californica:* a, foliage; b, fruit; (3) *Juglans ailanthifolia:* a, leaf; b, fruiting catkin; c, winter twig.

Juglandaceae   Walnut Family

Trees with thick twigs; large, alternate, pinnately compound leaves; and prominent leaf scars. Flowers of 2 kinds, both in catkins in ours, individually small, green, with adherent small bracts and bractlets, 4 small sepals and no petals; ovary adherent to the calyx. Fruit a drupe, with a more or less fleshy husk enclosing a hard endocarp (the nut) that contains a single oily seed.

Juglans   Walnuts

Trees with coarse twigs with chambered pith and large triangular leaf scars; the buds with few or no bud scales, the axillary ones sometimes doubled. Staminate flowers in axillary hanging catkins, with 8-40 stamens. Pistillate flowers in terminal spikelike catkins that become pendulous as the fruit develops; each flower formed of a single-chambered, inferior ovary crowned by a pair of stigmas. Fruit a drupe usually with a persistent husk.

Juglans ailanthifolia Carriere   Japanese Walnut
(*J. sieboldiana* Maximowicz not Goerppert)

Tree up to 20 m tall, with grey, shallowly grooved bark; coarse twigs with a brownish downy surface and a fringe of hairs along the upper edge of the leaf scar, which has three crescent-shaped bundle scars; large, downy terminal buds and smaller, sometimes doubled axillary buds. **Leaves** 40-60 cm long, with 7-17 leaflets with rounded bases, acuminate tips and serrate margins and with brownish star-shaped hairs beneath. **Catkins:** pistillate catkins have up to 20 flowers. **Fruit** several drupes along a hanging peduncle, ovoid, with sticky hairs; the stone or nut thick, pointed, rough, about 3 cm long (35). Plate 22, figure 3.

Range: Native of eastern Asia, planted here and occasionally escaping, as at Agassiz (5, 7).

Betulaceae   Birch Family

Deciduous trees or shrubs. Leaves alternate, petioled, stipulate; veins usually straight and pinnate, margins usually deeply serrate. Flowers monoecious, in catkins, usually appearing before the leaves; the pistillate catkins solitary or clustered, usually short, scaly; styles 2-cleft. Fruit a 1-celled and 1-seeded nut or winged nutlet.

Key to the Genera of Betulaceae
1a   Fruit a nut enclosed in a long sheath of united bracts . . . . *Corylus*
1b   Fruit small, winged nutlets in the axils of scalelike bracts.
    2a   Pistillate catkins solitary or paired; bracts thin, 3-lobed, deciduous;
      stamens 2 . . . . *Betula*
    2b   Pistillate catkins in racemes; bracts thick, persistent; stamens 4
      ·*Alnus*

Betula   Birches

Trees or shrubs with bark usually smooth and marked by numerous horizontally elongated lenticels. Winter buds ovoid, sharp-pointed in some species, and covered by several scales; terminal bud absent. Leaves alternate, simple; ovate, deltoid or circular; serrate, gland-dotted beneath, with 3-7 pairs of veins in ours; stipules papery. Catkins are composed of a series of flower groups, each group of 3 subtended by a bract and 3 bractlets in the staminate catkins and by a 3-lobed bract in the pistillate catkins. Staminate catkins long, cylindrical, pendent, clustered; pistillate catkins usually solitary, narrow, cylindrical. Staminate flowers with 2 bifid stamens; pistillate flowers are composed of the minute ovary with a forked style and linear stigmas. Fruit a minute, winged nutlet or samara, with remnant of the stigmas at apex. Fruiting catkin disintegrates, shedding bracts and nutlets.

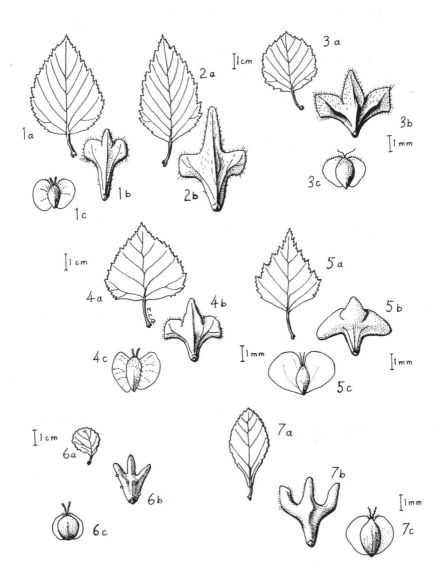

Plate 23 Birches. Leaves (a) to a common scale; bracts (b) and fruits (c) to a common scale. Figures: (1) *Betula papyrifera* var. *papyrifera;* (2) *B. papyrifera* var. *commutata;* (3) *B. occidentalis;* (4) *B. neoalaskana;* (5) *B. pendula;* (6) *B. glandulosa;* (7) *B. pumila.*

Key to the Species of *Betula*

1a Shrubs; leaves usually less than 25 mm wide, simply crenate or serrate, distinctly reticulate-veiny; fruit wings narrower than the central nutlet.

  2a Twigs densely resinous-glandular and thinly puberulent; leaves circular, with 10 or fewer teeth each side; samara wing less than half as wide as nutlet . . . . ***B. glandulosa***

  2b Twigs distinctly pubescent and sparsely glandular; leaves obovate and cuneate, with more than 10 teeth each side; samara wing at least half as wide as nutlet . . . . ***B. pumila***

1b Trees or large shrubs; leaves usually more than 25 mm wide, more or less biserrate, indistinctly reticulate-veiny; samara wings usually wider than nutlet.

  3a Bark becoming thick, furrowed and dark grey on trunk; leaves rhombic; pistillate bracts with recurving lateral lobes . . . . ***B. pendula***

  3b Bark remaining permanently thin and papery; leaves diverse; pistillate bracts with lateral lobes divergent but not recurved.

    4a Bark separable into thin layers, brownish to white; petioles more than 15 mm long.

      5a Twigs pubescent; leaves broadly ovate to rhombic, with tufts of hair in the vein axils beneath . . . . ***B. papyrifera***

      5b Twigs glabrous but glandular; leaves broadly deltoid, without axillary hair tufts beneath . . . . ***B. neoalaskana***

    4b Bark not separable into layers, reddish brown to black; petioles 5-15 mm long . . . . ***B. occidentalis***

Hybrids have been reported among all the species found in British Columbia (5, 17), including the exotics. Only the most commonly found hybrids are mentioned here.

Betula papyrifera Marshall   **Paper Birch**
White Birch, Canoe Birch

Tree up to 20 m tall and 60 cm in trunk diameter, rarely up to 35 m tall and 120 cm in trunk diameter (27, 31). Bark reddish brown or grey to white, and orange to pinkish beneath the outer surface; the outer bark can be easily peeled in sheets. Young trees have brown bark, as have the younger branches. Young branchlets loosely covered with long hairs and with scattered glands. **Leaves** 5-10 cm long, broadly ovate to rhombic, acuminate at apex, biserrate; glandular and hairy beneath, becoming glabrous except for tufts of hair in the axils of veins beneath; green on both sides. **Fruiting Catkins** hanging in pairs, 3-4 cm long; bracts 3-

lobed, ciliate, 5-6 mm long. **Fruit** a samara with wings slightly wider than the elliptic to obovate nutlet. Plate 23, figures 1 and 2.

Two varieties occur in British Columbia. The transcontinental var. *papyrifera* has broadly ovate leaves with rounded bases, bracts with short terminal lobes and ascending lateral lobes and white bark as an adult. The mainly western var. *commutata* (Regel) Fernald, the Western White Birch, has rhombic leaves obtuse or tapering at base and bracts with long terminal lobes and short, widely diverging lateral lobes; it often has grey or brown bark as an adult tree.

Range: Alaska to Newfoundland through the Boreal Forest Region, generally on low ground and southward to Colorado in the mountains. It occurs throughout the mainland of British Columbia, but is generally absent from the offshore islands. Rare on Vancouver Island, but reported in Saanich, near Victoria, where it was perhaps introduced.

Note: *Betula papyrifera* forms a complex swarm of hybrids with other species (5, 18, 29). Hybrids between *B. papyrifera* and *B. occidentalis* are called *B.* × *piperi* Britton and those between *B. papyrifera* and *B. neoalaskana* are called *B.* × *winteri* Dugle.

Betula occidentalis Hooker   Water Birch
(*B. fontinalis* Sargent)

Small tree, or more often a big shrub, forming clumps 4-10 m tall. Bark shiny reddish brown to blackish, not peeling readily. Branchlets slender, often somewhat pendulous, resinous-glandular, usually glabrous. **Leaves** 2.5-5 cm long, or longer on young shoots, broadly ovate or oval, acute at apex, rounded or cuneate at base; commonly biserrate, dark green above, paler green beneath, resinous gland-dotted on both surfaces. Petioles 1.5 cm long or less. **Fruiting Catkins** usually solitary, spreading, seldom hanging; bracts glabrous or shortly puberulent, lateral lobes widely divergent, angular, nearly equalling the narrow, acute middle lobe. **Fruit** a samara: a broadly elliptic nutlet, minutely hairy at the top, with wings about as wide as the nutlet. Plate 23, figure 3.

Range: From Alaska to Manitoba and southward to California, along streams and around lakes, often where slightly saline. In British Columbia, common east of the Coast Range; especially abundant in the drier parts of the southern interior, but absent from the coast.

Betula neoalaskana Sargent
Alaska Birch
Alaska Paper Birch

A small tree up to 10 m tall, with brownish bark becoming white on the trunk, flaky, but not as readily peelable as that of *P. papyrifera*. Branchlets glabrous or nearly so, densely resinous-glandular. **Leaves** deltoid, with acuminate apex, truncate at base, 2.5-5 cm long, the margin biserrate, glabrous or nearly so; the petiole 1.5-3 cm long. **Fruiting Catkins** solitary, hanging, 3-4 cm long; bracts ciliate, the widely divergent lateral lobes shorter than the acute central lobe. **Fruit** a samara with high-shouldered wings that are longer and wider than the nutlet. Plate 23, figure 4.

Range: Alaska to northwestern Ontario in the Boreal Forest Region; entering northeastern British Columbia, east of the Rocky Mountains; found in colder, wetter sites than *B. papyrifera*.

Betula pendula Roth
European Weeping Birch

Tree with initially white bark becoming thick, deeply fissured and dark grey. Branch-ends often, but not always, pendulous. Twigs finely glandular-roughened. **Leaves** rhombic, angular, obtuse at base and acute at tip, serrate, glabrous. Bracts of pistillate flowers with short terminal lobes and widely spreading and recurving lateral lobes. **Fruit** a samara with high-shouldered wings. Plate 23, figure 5.

Range: This European relative of our *B. neoalaskana* is our commonest cultivated birch; now escaped and becoming abundant on the Lower Mainland and to a lesser extent on southern Vancouver Island; now the commonest birch on the Fraser River delta.

Betula glandulosa Michaux
Dwarf Birch

Shrub up to 2 m tall, with densely glandular and thinly puberulent branchlets. **Leaves** circular to broadly elliptic, 10-25 mm long, rounded at apex, crenate-dentate with 5-9 teeth each side; the petiole very short. **Fruiting Catkins** erect, 1-2.5 cm long; bracts with nearly equal, divergent lobes rounded at apex. **Fruit** a samara with wings that are less than half as wide as the nutlet. Plate 23, figure 6. Photo 26.

Range: Transcontinental, in open areas with impeded drainage and in bogs. Widespread in the interior of British Columbia.

Betula pumila L. var. *glandulifera* Regel   **Swamp Birch**

Shrub up to 3 m tall, with slender, brown, puberulent and glandular branchlets. **Leaves** obovate-elliptic, up to 5 cm long by 3 cm wide, cuneate at base and acute at apex, crenate-serrate with 10 or more teeth each side, shiny above, pubescent on veins beneath when young. **Fruiting Catkins** erect, 1.5-2 cm long; bracts with upturned lateral lobes shorter than the terminal lobe, slightly hairy. **Fruit** a samara with wings ½ to ¾ as wide as the nutlet. Plate 23, figure 7.

Range: Yukon to Newfoundland in the Boreal Forest Region, in bogs and swamps, especially where calcareous, at low to subalpine levels. Widespread in British Columbia east of the Coast range, but seldom abundant and absent from northwestern British Columbia. The tall (to 4 m) form with more densely puberulent twigs found on Vancouver Island is forma *hallii* (Howell) Brayshaw.

Alnus   **Alders**

Trees and shrubs with smoothish, grey to brown bark; twigs with 2- or 3-angled pith, the buds often stalked, the terminal bud present. Leaves simple, pinnately veined. Catkins in clusters, hanging at flowering time. Staminate catkins in terminal clusters; the flowers in bracted groups, each flower with a 3-5-lobed calyx and 3-5 undivided stamens. Pistillate catkins in lateral clusters; the flowers in pairs subtended by completely joined, compound bracts; each flower consisting of an ovary with 2 linear stigmas but no calyx. Fruiting catkins short, ovoid, woody and conelike, persisting on the tree after seed fall. Fruit a small flat nutlet, winged or not.

Key to the Species of *Alnus*
1a Buds sessile; pistillate peduncles slender, longer than the cones; leaves shiny; nutlets with a broad wing . . . . *A. crispa*
1b Buds stalked; pistillate peduncles shorter than the cones; leaves dull, nutlets with a narrow wing or merely margined.
  2a Leaves glaucous and glabrous beneath, the margins revolute; nutlets with narrow membranous wings . . . . *A. rubra*
  2b Leaves green and pubescent beneath, margins not or slightly revolute; nutlets merely margined, without membranous wings
    . . . . *A. tenuifolia*

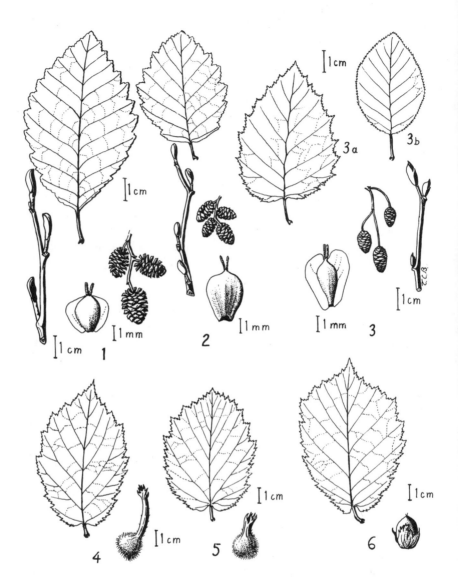

Plate 24 Alders and Hazels. Alders: Leaves, twigs, and fruiting catkins to a common scale; fruits to a common scale. Figures: (1) *Alnus rubra;* (2) *A. tenuifolia;* (3) *A. crispa:* 3a, leaf of subsp. *sinuata;* 3b, leaf of subsp. *crispa.* Hazels: Leaves and fruits in husks, to a common scale. Figures: (4) *Corylus cornuta* var. *cornuta;* (5) *C. cornuta* var. *californica;* (6) *C. avellana.*

Alnus crispa (Aiton) Pursh   Green and Sitka Alders

Shrub or small tree up to 8 m tall, with yellowish brown, scaly bark and sessile, pointed, axillary buds with 2 or 3 scales. The terminal bud may be stalked. **Leaves** 2-ranked, finely and sharply toothed, slightly gummy beneath when young, green and glabrous or puberulent beneath. **Catkins:** staminate catkins sessile, up to 10 cm long when flowering; pistillate catkins peduncled, cylindrical at flowering time; fruiting catkins ovoid, on slender peduncles at least as long as their catkins. **Fruit** a samara with membranous wings as wide as the nutlet. Plate 24, figure 3. Photo 21.

Key to the Subspecies of *A. crispa*
1a Leaf ovate to oval, simply dentate, the base rounded to cuneate;
    always a shrub . . . . **subsp.** *crispa*
1b Leaf broadly ovate, sinuate-margined as well as dentate, the base
    truncate to subcordate; big, coarse shrub or small tree . . . **subsp.** *sinuata*

Range: Subsp. *crispa,* the Green Alder, ranges from central Alaska to Newfoundland in the Boreal Forest Region. In British Columbia, confined to the northern and northeastern Boreal Forest Region. Subsp. *sinuata* (Regel) Hulten, the Sitka Alder, ranges from southern Alaska and the Aleutian Islands to California and Colorado and all over British Columbia. This species prefers better-drained, rockier soils than our other species of *Alnus*.

Alnus rubra Bongard   Red Alder

Tree up to 20 m tall and 1 m in trunk diameter, with smooth, grey bark; branchlets glabrous or slightly downy, dark red-brown, 3-angled in section with 3-angled pith. Leaf buds stalked, club-shaped, blunt, dark reddish brown. **Leaves** 7-15 cm long, ovate to oval; acute at apex, cuneate at base; bluntly biserrate, with rounded secondary teeth; the margin tightly revolute, glabrous or sparsely hairy, and glaucous beneath. Flowering before leaf expansion. **Fruiting Catkins** 4-8, up to 2.5 cm long, ovoid, on stout peduncles shorter than the catkins. **Fruit** a nutlet with narrow but distinct membranous wings. Plate 24, figure 1.

Range: Southern Alaska to California along the coast, and in Idaho; in moist soils from the coast into the valleys of the Coast and Cascade mountains. Reports of this species from the interior of British Columbia have not been confirmed.

A rare variety, var. *pinnatisecta* Starker, with pinnately lobed leaves with 5-7 pairs of oval lobes, has been found at Cowichan Lake and in the Nimpkish Valley.

Alnus tenuifolia Nuttall   Thinleaf Alder, Mountain Alder

Shrub or small tree up to 10 m tall, with greyish brown bark marked with horizontally elongated lenticels; branchlets brown, with yellowish brown tomentum, or glabrous; buds stalked, club-shaped. **Leaves** ovate; obtuse or acute at tip, rounded at base, bluntly biserrate, green and often yellowish puberulent beneath with a yellowish midvein. Flowering before leaf expansion. **Fruiting Catkins** conelike, ovoid, up to 17 mm long, on shorter peduncles. **Fruit** an obovate nutlet with an indistinct, narrow, thick border. Plate 24, figure 2.

Range: Alaska to Northwest Territories and southward in the mountains to Arizona, in moist alluvial sites and along streams. In British Columbia, widespread in the interior; generally absent from the coast, but present at Glen Lake, near Victoria (7).

Corylus   Hazels

Shrubs or small trees with yellowish hairy twigs; the terminal bud absent. Leaves alternate, simple, thin, strongly biserrate and slightly cordate at base. Leaf scars with 3 to several bundle scars. Flowers minute, appearing before the leaves in our species, in catkins. Staminate catkins arise from lateral buds and become elongate, loose and hanging at flowering time. The staminate flower consists of 4 divided stamens but no calyx, attached to 3 united, hair-tipped bracts. The pistillate flowers are in a bud adjacent to the twig end and consist of an ovary tipped by a vestigial calyx and 2 red stigmas and surrounded by a pair of united bracts. Only the stigmas emerge from under the bud scales. Fruit is a nut, broadly ovoid in ours, surrounded by an involucre of united bracts.

Key to the Species of *Corylus*

1a Envelope around nut deeply lobed, wide-mouthed, no longer than the nut, softly downy .... *C. avellana*

1b Envelope around the nut flask-shaped, prolonged in a tubular beak beyond the nut, bristly-hairy .... *C. cornuta*

Corylus avellana L.   Hazelnut

Big, coarse shrub up to 5 m tall, with pale brown glandular twigs and dark brown, smooth bark with horizontally elongated lenticels. **Leaves** broadly obovate to nearly circular, 5-10 cm long, with cordate base, abruptly acuminate tip and coarsely biserrate margins. **Catkins:** staminate catkins 3-4 cm long in winter, up to 10 cm long for early spring flowering; the bracts with fine, dull greyish puberulence. **Fruit** a softly puberulent nut, enclosed but not concealed by a puberulent involucre with lobes that appear ragged or torn. Plate 24, figure 6.

Range: Native of Europe; widely cultivated and escaped locally, in moist sites in the lower Fraser River valley and on Saltspring Island.

Corylus cornuta Marshall   Beaked Hazelnut
(*C. rostrata* Aiton)

Shrubs of varying habit, up to 4 m tall. Twigs pubescent, yellowish brown, with pubescent buds. **Leaves** elliptic to obovate, cordate to rounded at base, acuminate at apex, the margins biserrate. **Catkins** flowering before leaf expansion; staminate catkins yellowish and greyish-hairy, in flower hanging 4-7 cm; pistillate flowers visible as minute red stigmas emerging among the scales of buds. **Fruit** a globose nut completely enclosed in a flasklike, bristly involucre. Plate 24, figures 4 and 5.

Key to the Varieties of *C. cornuta*

1a Involucre with beak twice as long as nut; twig thinly pubescent, the hairs lasting only the first year; habit clumped or often colonial, scattered shoots arising from widely spreading rhizomes ..... **var.** *cornuta*

1b Involucre with a beak little if at all longer than the nut; twig more densely and persistently pubescent and often glandular; habit a big but compact clump .... **var.** *californica*

Range: Var. *cornuta* ranges from the interior of British Columbia northward at least to the Peace River basin, eastward to the Atlantic coast through the Boreal Forest Region, in wooded situations; absent from the coast of British Columbia. Var. *californica* (DC.) Sharp is the common variety on the coast, including southern Vancouver Island and recurs in moister parts of the Columbia Forest Region.

Fagaceae   Beech Family

Monoecious trees or shrubs with alternate, simple, pinnately veined leaves
and deciduous stipules. Flowers without petals; the staminate flowers in
catkins, containing up to 40 stamens; the pistillate flowers in heads,
catkins or solitary in leaf axils, surrounded by many-bracted involucres
and consisting of a 3-chambered ovary crowned by a vestigial calyx and 3
stigmas; containing up to 6 ovules, normally only 1 of which will become
an embryo or seed. Fruit a 1-chambered, 1-seeded nut in an involucre of
many united bracts.

Quercus   Oaks

Deciduous or evergreen trees or shrubs. Twigs with 5-angled pith.
Terminal bud present and surrounded by a cluster of similar axillary buds.
Bud with many scales, the lowest scale not directly above the leaf scar.
Leaf scar with several scattered bundle scars. Leaves alternate, 5-ranked,
tending to be crowded toward the twig end; often pinnately lobed or
toothed, the main lateral veins running into the lobes. Stipules present
briefly. Staminate flowers in slender catkins, each flower with 3-12 stamens in a 3- to 8-lobed calyx. Pistillate flowers in many-bracted involucres that are solitary in the leaf axils or in catkins; each flower a
3-chambered ovary crowned by the vestigial calyx and 3 stigmas. Fruit a
nut (the acorn) seated in a scaly, cuplike involucre.

Key to the Species of *Quercus*
1a Leaf lobes blunt to rounded; twigs and buds hairy; native, coastal
.... *Q. garryana*
1b Leaf lobes acute and bristle-tipped; twigs and buds glabrous; introduced inland .... *Q. rubra*

Quercus garryana Douglas *ex* Hooker   Garry Oak

Small to large trees, or even shrubs, depending on environment; up to 25
m tall by 150 cm in diameter in the Victoria area; trunk short below the
lowest branches, and a large, round-topped crown; rough, greyish bark

with narrow, shallow furrows in a rectangular pattern. Young branchlets and the clustered large buds covered with pale brownish hairs. **Leaves** obovate, 10-15 cm or more long, cuneate to rounded at base, sinuate with rounded or truncate lobes; dark green and glabrous above, paler green and minutely hairy beneath; stiff and rather leathery in texture when mature; the margin sometimes slightly revolute. Petiole stout, 12-20 mm long. **Fruit** solitary or sometimes paired. Acorn ovoid, about 25 mm long, light brown, maturing in its first year; seated in a hard, shallow cup with hairy scales free at their tips. Plate 25, figure 1. Photo 25.

Range: Southwestern British Columbia to California from the coast to the Cascade Mountains, and in the Sierra Nevada, avoiding the outer coast. In British Columbia, along the east coast of Vancouver Island from Sooke to near Courtenay and at Port Alberni; also at two locations in the Lower Fraser Valley: on Sumas Mountain and at Yale. It prefers dry, open sites, demanding full sunlight and freedom from conifer competition.

Quercus robur L., English Oak, which is frequently planted, has been found wild in the Vancouver area and at Yale (Gerald B. Straley, personal communication). It resembles Garry Oak, but its sinuately-lobed leaves have basal recurved lobes (auricles) and its acorns are on long peduncles.

Quercus rubra L.   Red Oak
(*Q. borealis* Michaux)

Tree up to 30 m tall with smooth, dark grey bark becoming furrowed into hard ridges in age. Twigs dark reddish brown, smooth, glabrous or almost so. **Leaves** pinnately lobed, with V-shaped sinuses and acuminate, bristle-tipped lobes; dull, glabrous, 10-17 cm long. **Flowers:** staminate flowers in slender catkins; pistillate flowers 1 or 2 together, almost sessile on a twig. **Fruit** ripening in its second year, in a shallow glabrous cup with closely overlapping smooth, brown scales. Plate 25, figure 2.

Range: Widespread in eastern North America, on dry, rocky sites; widely planted elsewhere. In British Columbia, established wild at Revelstoke (7, 61).

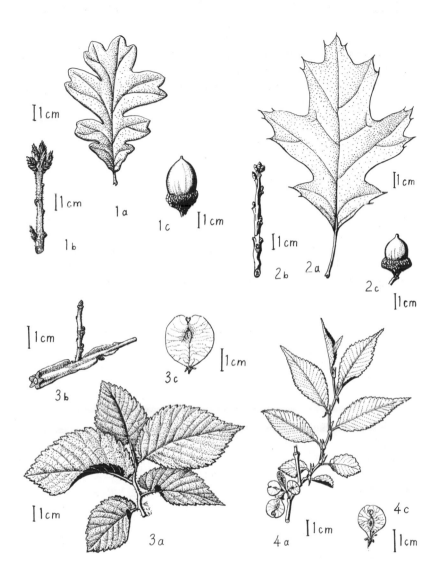

Plate 25 Oaks and Elms. Figures: (1) *Quercus garryana;* (2) *Q. rubra;* (3) *Ulmus procera;* (4) *U. pumila.* a, leaves and branchlets; b, winter twigs; c, fruit.

Ulmaceae   Elm Family

Trees and shrubs with alternate, stipulate, usually obliquely-based, simple, 2-ranked leaves. Flowers on axillary, clustered pedicels; small, without petals, with 4-8 basally connate sepals; the same number of stamens joined to the sepal bases and a superior ovary consisting of 2 carpels united to form a single locule containing a single erect ovule. Style short, stigmas 2. Fruit a samara or drupe.

Ulmus   Elms

Tree with distinctly 2-ranked branching and twigs with leaf scars with 3 bundle scars. Buds with alternate 2-ranked bud scales; the terminal bud absent. Leaves alternate, 2-ranked, asymmetrical; the left and right leaves mirror images of each other; pinnately veined. Flowers and fruit appearing before the leaves in our species. Fruit a samara formed of a central nutlet surrounded by a disklike wing often apically notched and cupped at base by the remnants of the calyx.

A genus of about 30 species of north temperate latitudes; none native west of the Rocky Mountains in North America, but 2 species naturalized in British Columbia.

Key to the Species of *Ulmus* in British Columbia
1a Leaves almost symmetrical at base, simply serrate, smooth-surfaced; bark very pale grey . . . . *U. pumila*
1b Leaves distinctly asymmetrical, biserrate, rough-surfaced; bark brown *U. procera*

Ulmus procera Salisbury   English Elm
(*U. campestris* Miller)

Tree up to 30 m tall, commonly producing abundant suckers. Branchlets pubescent when young, the light brown bark developing longitudinal corky flanges. **Leaves** very oblique at base; ovate and very rough above; soft-pubescent with axillary tufts of hair below; biserrate; 50-75 mm long; petioles 4-6 mm, pubescent. **Flowers** densely clustered in fascicles on very short pedicels that are not pendulous; stamens 3-5; stigmas white, calyx with 4-7 lobes. **Fruit** orbicular to broadly obovate; glabrous; seed near the apex, which has a small, closed notch. Plate 25, figure 3.

Range: Native of Europe, commonly planted. Escaping around Victoria and producing dense thickets by suckering.

Ulmus pumila L.   Siberian Elm, Chinese Elm

Tree up to 25 m tall but commonly smaller, with long slender branches and branchlets with pale grey to grey-brown bark; twigs glabrous or hairy. **Leaves** alternate, 2-ranked, elliptic, almost symmetrical, simply serrate, 2-7 cm long; cuneate at base, acute to acuminate at apex; smooth. **Flowers** before leaf development in spring. **Fruit** in dense clusters at nodes of the previous year's twig; circular samara about 1 cm across, slightly notched at apex, on pedicels 2-3 mm long. Plate 25, figure 4.

Range: Native to Siberia, China and Mongolia; commonly planted in arid regions and escaping: known wild in the Okanagan, Similkameen and Kettle valleys.

Moraceae   Mulberry Family

Trees, shrubs, vines or herbs with milky juice. Flowers minute, of 2 kinds in densely crowded heads, spikes or catkins. Calyx 4-lobed, or sometimes absent from pistillate flowers; no petals; stamens as many as calyx lobes and opposite them, or fewer. Ovary of 2 carpels, with 2 stigmas and 1 ovule. Fruit is a multiple one, formed from the aggregation of flowers in a spike or head.

Morus   Mulberries

Deciduous trees or shrubs with alternate, 2-ranked leaves, early-deciduous stipules and milky juice. Staminate and pistillate flowers in separate, axillary spikes or catkins. Fruit a fleshy blackberry-like aggregate formed of the fleshy spike of pistillate flowers, the achenes that are formed from the ovaries being embedded in the fleshy calyces.

Morus alba L.   White Mulberry

Tree up to 15 m tall, rather crooked and spreading in form, with yellowish wood and rough, greyish bark. Twigs with 2-ranked leaf scars, each with several bundle scars in a crescent or ellipse. Buds with 2-ranked bud scales, the lowest scales sometimes subtending secondary buds. Terminal bud absent. **Leaves** highly variable in shape, generally ovate, commonly asymmetrical; pinnately lobed or unlobed; simply serrate except for the sinuses between lobes, with three strong veins from the base; petioles and undersides of the main veins puberulent. **Flowers** in axillary catkins; the staminate in drooping catkins about 25 mm long, the pistillate in short, dense, spreading or drooping catkins 10-15 mm long. **Fruit** passing from white through red and purple to almost black; succulent and sweet. Plate 26, figure 1.

Range: Native of Asia, long cultivated and now widely naturalized in North America; long established on riverbanks around Spences Bridge, British Columbia.

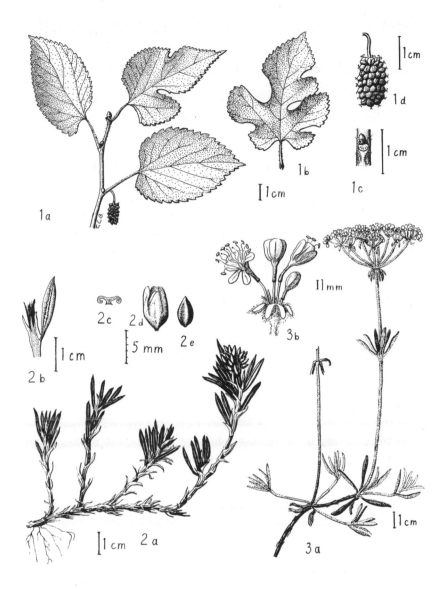

Plate 26 Figures: (1) *Morus alba:* 1a, twig with foliage and fruit; 1b, deeply lobed leaf form; 1c, leaf scar and axillary bud; 1d, fruit; (2) *Polygonum paronychia:* 2a, stem, foliage, and inflorescence; 2b, leaf (dorsal view) and ocrea; 2c, cross section of leaf; 2d, flower; 2e, nutlet; (3) *Eriogonum heracleioides:* 3a, branch with inflorescence; 3b, ultimate umbellet and flowers.

Polygonaceae   Knotweed Family

A mainly herbaceous family (at least in temperate latitudes) with perfect flowers lacking petals. Sepals 3-6, joined into a commonly funnel-shaped calyx. Stamens 8 or 9, inserted on the inner surface of the calyx; usually a tricarpellate ovary with 3 styles (or, by union, 1). Fruit in ours a 3-angled nutlet containing a single seed.

Key to the Woody Genera of Polygonaceae in British Columbia
1a   Leaves alternate but often crowded, each with a stipular sheath (called an *ocrea*) surrounding the stem. Flowers axillary or in spikes. Sepals 5, stamens 8. Ours a prostrate shrublet .... *Polygonum*
1b   Leaves in whorls in ours, without ocreas. Flowers in compound involucrate umbels. Sepals 6, stamens 9. Prostrate to ascending woody stems giving rise to erect flowering stems .... *Eriogonum*

Eriogonum   **Wild Buckwheat, Umbrella Plant**

Herbaceous perennials, annuals or subshrubs; the stems and leaves commonly grey with tomentum. Leaves often whorled, but alternate in some; simple, ovate to lanceolate. Flowers on jointed pedicels clustered in small funnel-like involucres; the involucres variously axillary, cymosely or umbellately arranged; the umbels subtended by whorls of bracts. Sepals united to various degrees. Stamens 8, styles 3, separate.

Eriogonum heracleioides **Nuttall**   **Umbrella Plant**
var. *angustifolium* **(Nuttall)**   **Parsnip-flowered Eriogonum**
Torrey & Gray

Low, sprawling, woody-based subshrub, generally of a pale greyish aspect. The lower, permanent stems prostrate or oblique to the ground; woody, dark grey, usually less than 20 cm high. **Leaves** whorled, lanceolate to almost linear in ours; 2-6 cm long, 3-5 mm wide; short-petioled, greyish-tomentose. **Flowering branches** erect, up to 20 cm long, with a whorl of leaves about midway up; pale greyish-woolly; terminating in a compound umbel subtended by a whorl (involucre) of bracts; the constituent umbellets similarly subtended. Involucres of the flowers bell-shaped, greyish-woolly, the many lobes reflexed. Calyx glabrous, white to cream-coloured, occasionally with an orange or pinkish cast; the lobes (sepals) divided almost to the base. Plate 26, figure 3. Photo 28.

Range: Interior of British Columbia to California, Montana and the Great Basin states. Commonly found on sandy or gravelly soils of semi-arid grassland, on open dry parkland or on deserts farther south.

Polygonum   Knotweed

Mostly herbaceous plants, but including a few shrubs and vines; with alternate, simple leaves; the stipules united into tubular or flaring sheaths surrounding the stem. Flowers few in leaf axils or aggregated in terminal spikes. Calyx 5-lobed; stamens 8. Fruit a 3-angled nutlet, resembling that of Buckwheat.

Polygonum paronychia   **Black Knotweed**
Chamisso & Schlechtendal

Prostrate subshrub; the creeping stems covered with the torn remains of old stipules. **Leaves** closely spaced toward the branch tips; narrowly lanceolate to nearly linear, jointed at base, revolute-margined, 1-3 cm long; the midrib ciliate beneath. **Flowers** clustered in the upper leaf axils; short-stalked or almost sessile. Calyx white to pink, 5-6 mm long, the lobes (sepals) united for ¼ to ½ their length. Stamens 8, included within the calyx. Styles 3, united more than half their length, 1.5 mm long. **Fruit** a nutlet 4-5 mm long; black, smooth and shiny. Plate 26, figure 2. Photo 27.

Range: Vancouver Island to California. Strictly coastal, on sand dunes or rocks above the shore.

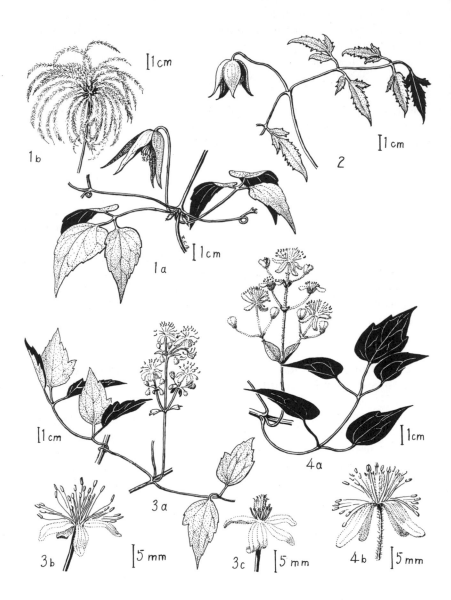

Plate 27 Clematis. Figures: (1) *Clematis occidentalis:* 1a, leaves and flower; 1b, head of achenes; (2) *C. tangutica;* (3) *C. ligusticifolia:* 3a, leaf and inflorescence; 3b, staminate flower; 3c, pistillate flower; (4) *C. vitalba:* 4a, leaf and inflorescence; 4b, flower.

Ranunculaceae   Buttercup Family

Mainly herbs, rarely woody vines or shrubs. Leaves alternate or opposite, simple or compound, often with sheathing bases, but usually without distinct stipules. Flowers radially or bilaterally symmetrical, with all parts separate; sepals 3-15, often petal-like; petals often the same number as sepals, but sometimes reduced to nectaries or absent; stamens 10 to many; carpels 1 to many, separate, containing one or more ovules, in fruit becoming follicles or achenes.

Clematis   Clematis

Vines. Leaves opposite, compound. Flowers perfect or dioecious, radially symmetrical, with 4 or 5 or more petal-like sepals; petals none; stamens many; the numerous carpels becoming achenes with long, plumose styles (6).

Key to the Species of *Clematis*
1a Leaves with 3 leaflets; flowers solitary, large, blue . . . *C. occidentalis*
1b Leaves pinnately compound, with 5 or more leaflets.
   2a Flowers solitary, yellow . . . . *C. tangutica*
   2b Flowers in panicles; small, white.
      3a Flowers dioecious; either staminate or pistillate, on different
         plants . . . . *C. ligusticifolia*
      3b Flowers perfect . . . . *C. vitalba*

Clematis occidentalis (Hornemann) DC.   **Blue Clematis**
var. *grosseserrata* (Rydberg) Pringle
(*C. columbiana* [Nuttall] Torrey & Gray;
C. verticillaris DC. var. *columbiana* [Nuttall] Gray)

Climbing or trailing vine, 1-4 m long. **Leaves** trifoliolate; leaflets ovate, acuminate, rounded at base, the lateral leaflets asymmetrical. **Flowers** solitary, on long pedicels terminating short axillary shoots, with violet-blue sepals. **Fruit** an achene with a plumose tail 4-5 cm long. Plate 27, figure 1. Photo 22.

Range: Var. *grosseserrata* from Alberta and British Columbia southward to Colorado. In British Columbia, found in coniferous forests at moderate elevations across the southern interior, northward to the Peace River. Var. *occidentalis* is in eastern Canada and the United States, but not in British Columbia.

Clematis ligusticifolia Nuttall in Torrey & Gray

White Clematis
Old-man's-beard

Climbing vine to 10 m. **Leaves** pinnate, usually 5-foliolate, the lowest pair of leaflets occasionally trifoliolate. Leaflets ovate to ovate-lanceolate, 3-10 cm long, acute or acuminate, rounded or subcordate, coarsely toothed or sometimes entire, occasionally 3-lobed, thin; petioles 1-3 cm. **Flowers** about 2 cm across, white, dioecious, in axillary or terminal clusters; sepals densely pubescent outside. **Fruit** a plumose achene. Plate 27, figure 3. Photo 30.

Range: British Columbia to California and the Dakotas. In British Columbia, across the dry southern interior, northward to Alkali Lake and Fairmont Hot Springs; found in open, sunny situations and deciduous woods.

Clematis vitalba L.

Traveller's Joy

Resembling *C. ligusticifolia,* climbing up to 10 m. **Leaves** pinnate, occasionally bipinnate; leaflets variously lobed or toothed, the teeth not turned outward. **Flowers** in panicles, white, perfect; sepals densely pubescent outside. Plate 27, figure 4.

Range: Native of Europe; escaping from cultivation in the Vancouver area, on southern Vancouver Island and on several smaller intervening islands. Sometimes mistaken for *C. ligusticifolia.*

Clematis tangutica (Maximowicz) Korshinsky

Climbing vine to 3 m high. **Leaves** pinnate or bipinnate; leaflets lanceolate with outwardly spreading teeth. **Flowers** solitary, nodding, 4-5 cm across, yellow, with 4 sepals. **Fruit** a large cluster of achenes with long, plumose styles. Plate 27, figure 2.

Range: Native of Central Asia; escaping from cultivation at widely scattered localities in central and eastern British Columbia, northward to the Peace River and westward to Ootsa Lake and also on Saltspring Island on the coast.

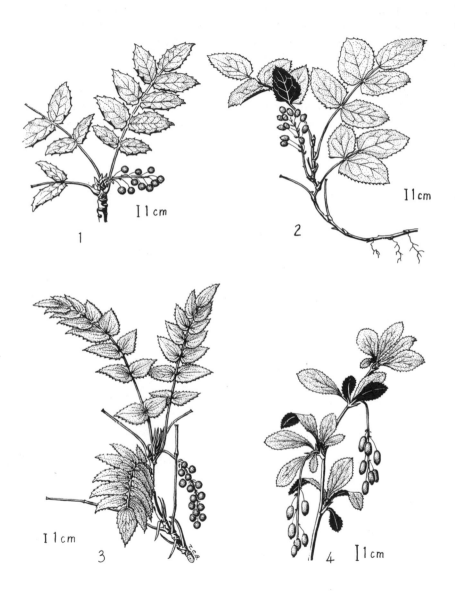

Plate 28 Barberries. Figures: (1) *Berberis aquifolium* subsp. *aquifolium;* (2) *B. aquifolium* subsp. *repens;* (3) *B. nervosa;* (4) *B. vulgaris.*

Berberidaceae   Barberry Family

Shrubs or herbs. Leaves alternate, simple or compound, evergreen or deciduous; stipulate, or the petiole dilated at base. Flowers perfect, variously arranged, in racemes or spikes in our species. Sepals and petals in whorls of 3 or sometimes 2 or 4, or occasionally absent. One stamen opposite each petal. Anthers usually opening by uplifting valves. Ovary superior, usually of a single carpel. Style short or none; the stigma usually button-shaped. Fruit a berry or pod.

Berberis   Barberries

Shrubs, often prickly, with yellow wood. Leaves alternate, of diverse form; compound in ours, with minute stipules. Flowers in racemes in our species; with parts in whorls of 3; sepals 6 or 9; petals 6, usually yellow. Anthers opening by uplifting valves. Fruit a berry formed from a single carpel bearing a buttonlike stigma and containing a few to several seeds.

    A genus of about 500 species, including most of the species in the family. The subgenus *Mahonia* is often treated as a separate genus.

Key to the Species of *Berberis* in British Columbia
1a   Leaves apparently simple (actually compound with a single leaflet), deciduous, in fascicles in the axils of branched spines (modified leaves) attached to angular stems . . . . **subg.** *Berberis; B. vulgaris*
1b   Leaves pinnately compound, evergreen, alternate on a cylindrical stem; the leaflets with spinulose marginal teeth . . . . **subg.** *Mahonia*
   2a   Leaflets palmately veined, dull-surfaced. Bud scales 2-4 cm long, persistent . . . . *B. nervosa*
   2b   Leaflets pinnately veined, often shiny. Bud scales less than 2 cm long, often deciduous or disintegrating in a year . . . . . *B. aquifolium*

Berberis aquifolium **Pursh**   **Tall Oregon-grape**
(*Mahonia aquifolium* **[Pursh] Nuttall**)

Rhizomatous shrub varying from a few centimetres to 4 m tall, but usually 1 or 2 m tall. Branches coarse, cylindrical in section. Bud scales triangular, less than 2 cm long, usually shed or disintegrating within a year or two. **Leaves** pinnately compound, with 5-11 leaflets. Leaflets leathery, evergreen, pinnately veined, often shiny above; the marginal teeth tipped with little spines. **Flowers** in erect racemes from the axils of bud scales, yellow, the stamens with small lateral appendages below the anthers. **Fruit** a blue-black berry with a greyish bloom, in a raceme that may become horizontally spreading. Plate 28, figures 1 and 2. Photo 29.

Key to the Subspecies and Varieties of *B. aquifolium*
1a Leaflets 3-7, less than twice as long as wide, often dull above; the margin with 10-40 finely spinulose teeth. Bush up to 0.5 m tall
.... **subsp.** *repens* **(Lindley) Brayshaw**
1b Leaflets 5-11, averaging more than twice as long as wide, usually shiny above; the margins with 8-20 spine-tipped teeth. Bush up to 4 m tall .... **subsp.** *aquifolium*
2a Leaflets coarsely toothed, the spinules 2-4 mm long
.... **var.** *aquifolium*
2b Leaflets serrulate, the spinules 1 mm or less long
.... **var.** *lyallii* **Ahrendt**

Range: British Columbia to Texas and from the Pacific coast to South Dakota, in the forest and forest edge, usually on coarse, well-drained soils. Across southern British Columbia from Vancouver Island to the Rocky Mountains, and northward to the Prince George area. Subsp. *repens* is confined to the interior, usually in dry, open situations (6).

Berberis nervosa **Pursh**   **Cascade Oregon-grape**
(*Mahonia nervosa* **[Pursh] Nuttall**)   **Low Oregon-grape**

Rhizomatous shrub, 30-40 (sometimes 75) cm tall. The stem with persistent, lanceolate bud scales 25 mm or more long. **Leaves** with 9-19 leaflets. Leaflets ovate, dull-surfaced, palmately veined, rigid, with oblique spine-tipped teeth. **Flowers** bright yellow, in initially erect racemes 10-20 cm long. **Fruiting Racemes** arching or hanging. Berry blue-black with a bloom. Plate 28, figure 3. Photo 31.

Range: British Columbia to California, from the coast to the Cascade Range. In British Columbia, in woodlands on well-drained soils; more shade-loving than *B. aquifolium*. Mainly coastal; Vancouver Island northward to Nimpkish Valley. On the mainland coast, where commoner than *B. aquifolium,* eastward to the lower Fraser Canyon and the Skagit Valley. Also in the southern Monashee and Selkirk Mountains, where scarce.

Berberis vulgaris L.   **Common Barberry**

Erect clump-forming shrub up to 2.5 m tall with somewhat arching branches, which are angled in cross section. **Leaves** on the long shoots are modified into usually 3-pronged spines. Normal leaves in fascicles or short shoots in the axils of the spines; the blades obovate, tapering to joints at the top of the very short, persistent petioles; serrulate, conspicuously net-veined beneath, 1-4 cm long. **Flowers** in arching or hanging racemes up to 6 cm long, arising in the axils of leaves. **Fruit** ellipsoid, red or purple. Plate 28, figure 4.

Range: Native of Europe; formerly cultivated and sparingly escaped on southern Vancouver Island, in the Vancouver area and in the Okanagan Valley northward at least as far as Vernon. Its cultivation is prohibited, since it serves as an alternate host of the stem rust disease of wheat.

Grossulariaceae   Gooseberry Family

Shrubs, usually glandular and often prickly. Leaves alternate, simple, palmately veined and lobed and usually serrate; without stipules, but with long gland-tipped hairs on the basal parts or the petioles. Flowers in racemes that are usually on short lateral branchlets; pedicels usually bracted. Ovary inferior, 1-celled, with many ovules; calyx cup extending beyond the top of the ovary to form a hypanthium. Sepals 5, coloured, the most conspicuous appendages of the flower. Petals 5, smaller than the sepals; they and the 5 stamens, which are opposite the sepals, inserted just within the top of the hypanthium. Styles basically 2, but united to varying degrees. Fruit a juicy berry, often glandular and sometimes bristly.

A widespread family formerly treated as a subfamily of the family Saxifragaceae, of about a dozen genera. Most of the 150 species belong to the genus *Ribes,* to which the above description applies.

Ribes   Gooseberries and Currants

The only genus of this family in North America, with over a hundred species; distributed over the temperate regions of the Northern Hemisphere and in the Andes and far southern temperate regions of South America.

Key to the Species of *Ribes*
1a Stems armed with spines at nodes and often with prickles.
   2a Flowers more than 4 on a drooping raceme; pedicels jointed below ovary; hypanthium saucer-shaped (**subg.** *Grossularioides,* **Bristly Currants**).
      3a Fruits purple-black; leaves mostly glabrous and not glandular; pedicels slender, 1.5-3 times as long as bracts . . . . *R. lacustre*
      3b Fruits red; leaves pubescent and slightly glandular; pedicels stout, usually less than twice as long as bracts . . . *R. montigenum*
   2b Flowers 1-4 per inflorescence; pedicels not jointed; hypanthium cup- shaped to tubular (**subg.** *Grossularia* (**genus** *Grossularia* **Miller): the Gooseberries).**
      4a Ovary and fruit covered with stalked glands . . . . *R. lobbii*
      4b Ovary and fruit smooth, non-glandular.

5a Hypanthium short, cup-shaped; little, if any, longer than wide.
   6a Stamens shorter than to barely equalling the sepals.
   Interior of British Columbia . . . . *R. irriguum*
   6b Stamens longer than sepals. Coastal . . . . . *R. divaricatum*
5b Hypanthium tubular, distinctly longer than wide.
   7a Underside of leaf and upper petiole usually with short-
   stalked glands; petals and stamens nearly equal, 2-2.5 mm
   long . . . . *R. oxyacanthoides*
   7b Underside of leaf and upper petiole glandless; occasional-
   ly long-stalked glands on lower petiole; petals 1-2 mm long;
   stamens 2-4 mm long, 1.5-2 or more times as long as petals
   . . . . *R. inerme*
1b Stems not prickly; racemes with 4 to many flowers; pedicels jointed
(**subg. *Ribes*, Currants**).
   8a Hypanthium small, saucer-shaped, wider than long.
      9a Ovary smooth, without hairs or glands; berries red
      . . . . *R. triste*
      9b Ovary glandular, hairy or both.
         10a Ovary with sessile yellow glands; berries black or bluish.
         11a At least the lower bracts longer than their axillary
         pedicels; flowers greenish. Coast to Cascade and Coast
         ranges . . . . *R. bracteosum*
         11b All bracts shorter than their pedicels; flowers white.
         Interior of British Columbia . . . . *R. hudsonianum*
         10b Ovary with stalked glands.
         12a Petals wider than long, half-moon-shaped; berries
         blue-black.
            13a Racemes drooping; stamens broadened and some-
            times united at their bases . . . . *R. howellii*
            13b Racemes spreading to erect; stamens slender, free
            . . . . *R. laxiflorum*
         12b Petals longer than wide, cuneate-spatulate; berries red
         . . . . *R. glandulosum*
   8b Hypanthium at least as long as wide, tubular to cup-shaped.
      14a Leaves distinctly 3-lobed; the narrow, rounded lobes sepa-
      rated by wide sinuses. Pedicel, ovary and tube quite glabrous;
      sepals about two-thirds the length of tube; flowers yellow
      . . . . *R. aureum*
      14b Leaves usually 5-lobed, the lobes serrate or almost
      unlobed. Pedicel, ovary and tube with stalked glands and some-
      times hairy; flowers not yellow.

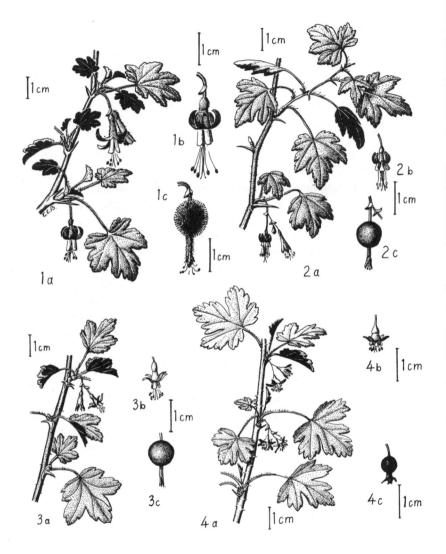

Plate 29 Gooseberries. Figures: (1) *Ribes lobbii;* (2) *R. divaricatum;* (3) *R. inerme;* (4) *R. irriguum.* a, flowering branch; b, flower; c, fruit.

15a Sepals longer than tube; flowers red; anthers not glandular at tips. Coastal .... *R. sanguineum*
15b Sepals shorter than or equalling the tube; flowers whitish or creamy; anthers with cup-shaped apical glands. Interior of British Columbia.

16a Sepals half or less as long as tube; berries red; leaves semicircular, shallowly notched and toothed, scarcely lobed .... *R. cereum*
16b Sepals about ¾ as long as tube; berries blue-black; leaves shallowly 5-lobed .... *R. viscosissimum*

Ribes lobbii Gray   Gummy Gooseberry

Shrub up to 2 m tall; branches bearing usually triple spines at the nodes, the internodes glandular but not bristly. **Leaves** usually less than 3 cm across, 5-lobed; cordate or subcordate at base, with blunt to round-tipped lobes; bluntly serrate, somewhat glandular-hairy, especially on the petioles. **Racemes** drooping, about 20 mm long, 1- to 3-flowered, the flowers about 20-25 mm long; ovary densely glandular with stalked glands; hypanthium tubular, finely hairy; sepals reflexed, red or purple; petals oblong, white; stamens exserted; anthers purple or reddish, warty on back. Styles as long as the stamens, glabrous, divided toward the tip. **Fruit** purple, densely glandular-bristly, 15 mm long. Plate 29, figure 1.

Range: Southwestern British Columbia to northwestern California, from the coast to the Cascade Range, in open, sunny woods; in British Columbia, restricted to southern Vancouver Island and islands in the Strait of Georgia. A natural hybrid, *R. divaricatum* × *lobbii,* has been found near Victoria.

Ribes divaricatum Douglas   Wild Gooseberry

Shrub 1.5-3 m tall; branches usually not very bristly; spines stout, usually simple, 1 cm or more long, often rather recurved. **Leaves** 2-3 cm wide, orbicular in outline; 3- to 5-lobed, with toothed margins; slightly hairy above, more densely so beneath; pedicels glabrous. **Flowers** 1-4; bracts short; sepals purplish or sometimes greenish, reflexed; petals cuneate, half as long as the sepals, erect, white. Stamens and styles longer than the sepals, well exserted; styles hairy, united at base. **Fruit** glabrous, black or purple, about 1 cm across. Plate 29, figure 2.

Range: Woods and shores, from the coast into the Coast and Cascade ranges, from southern British Columbia to California; in British Columbia, extending inland as far as Manning Provincial Park and Lytton.

Ribes inerme Rydberg

Black Gooseberry
White-stemmed Gooseberry

Shrub up to 2 m tall; branches with short nodal spines commonly 3-pronged, with or without a few short internodal bristles. **Leaves** up to 5 cm or more wide, truncate at base; lobes 3-5, obtuse; the margin crenate-dentate and finely ciliate. Petiole and leaf surface thinly tomentose, often becoming glabrous. **Flowers** 1-3 in a short raceme; hypanthium cup-shaped, longer than wide, glabrous; sepals oblong, pale green, reflexed; petals white, rarely pink, about half as long as the sepals; stamens about as long as the sepals; ovary glabrous; style as long as the sepals or a little longer, divided into two about halfway to the base, hairy below. **Fruit** purplish, 8 mm across, edible. Plate 29, figure 3.

Range: Southern British Columbia, from the Coast and Cascade ranges eastward to southwestern Alberta, northward to the Prince George area and southward to California.

Note: *Ribes inerme* is weakly distinguished from the related *R. irriguum.* Plants have been found that share, the "wrong" way around, the supposedly distinguishing characteristics of these two species, or have hypanthia of equal length and width.

Ribes irriguum Douglas

Idaho Gooseberry
Inland Black Gooseberry

Shrub up to 3 m tall, similar to *R. inerme.* Stems pale, with short, simple spines at the nodes and sparse, fine bristles on the internodes. **Leaves** 2-8 cm wide, cordate to truncate at base, pentagonal in outline; 3- to 5-lobed, the lobes with obtuse to rounded teeth; the blade sparingly hairy above, glabrous beneath; ciliate-margined; petioles often glandular and hairy above. **Flowers** 1-3 in very short racemes from short lateral branchlets; with glandular-ciliate bractlets shorter than the pedicels. Hypanthium bowl-shaped, at least as wide as long, glabrous. Sepals greenish white; petals white, spatulate, half or less as long as the sepals; the stamens about as long as the sepals. Ovary glabrous; style hairy below, divided less than halfway to the base. **Fruit** glabrous, black to bluish purple. Plate 29, figure 4.

Range: Southern interior of British Columbia to northeastern Oregon and western Montana. In British Columbia, northward to around Canal Flats and westward to Lytton.

Ribes lacustre (Persoon) Poiret in Lamarck   Bristly Currant

Shrub 1-1.5 m tall with ascending to spreading branches; twigs pale, hairy. STEM armed with 3- to 5-parted nodal spines and many short, fine internodal prickles. **Leaves** 3-8 cm across, pentagonal in outline, cordate at base, deeply 3- to 5-lobed, the lobes with blunt teeth; glabrous or sparsely hairy on both sides; petioles with long gland-tipped hairs and fine, short puberulence. **Racemes** axillary, drooping, often with a zigzag axis and 7-20 flowers; bracts oblong, thin, shorter than the pedicels, which are jointed just below the flowers; peduncle, pedicel and ovary with stalked glands. **Flowers** 5-7 mm wide; hypanthium saucer-shaped, glabrous; sepals more or less orbicular, reddish-tinged; petals pale, wider than long; they and the stamens shorter than the sepals. **Fruit** black, bristly-glandular, though often sparsely so. Plate 30, figure 1.

Range: Transcontinental-boreal: Alaska to Newfoundland and southward to California; general throughout British Columbia, commonly as an undershrub in moist coniferous or mixed woods.

Ribes montigenum McClatchie   Alpine Bristly Currant

Low shrub 0.5-1 m tall, resembling a small *R. lacustre,* but of more spreading habit. Stem densely bristly. **Leaves** 1-4 cm across, orbicular to pentagonal in outline, cordate at base; 5-lobed, the lobes with blunt teeth; finely hairy and sparsely glandular. **Racemes** short, 4- to 9-flowered, the peduncles and pedicels glandular-hairy; bracts about as long as the pedicels; pedicels jointed just below the flowers. **Flowers** 6-9 mm wide, similar to those of *R. lacustre;* sepals and petals pinkish, petals cuneate and truncate. **Fruit** red, glandular-bristly. Plate 30, figure 2.

Range: Southern British Columbia to California and New Mexico in rocky sites at subalpine or alpine levels. Recorded in extreme southern interior of British Columbia (Flathead, Natal, Manning Provincial Park); but all

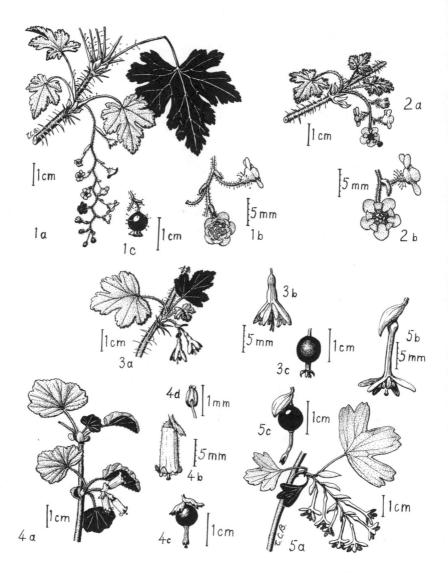

Plate 30 Gooseberries and Currants. Figures: (1) *Ribes lacustre;* (2) *R. montigenum;* (3) *R. oxyacanthoides;* (4) *R. cereum;* (5) *R. aureum.* a, flowering branches; b, flowers; c, fruit; d, anther with apical gland.

British Columbian specimens I have seen purporting to be *R. montigenum* are intermediate in character between it and *R. lacustre*, and could as readily be assigned to the latter species. The illustration is based on specimens of *R. montigenum* from Utah. Watch for more typical examples of this species at alpine levels in the southern interior of this province.

Ribes oxyacanthoides L. **Smooth Gooseberry**
Northern Gooseberry

Low shrub under 1 m tall; twigs finely puberulent, with divided nodal spines and usually with internodal prickles. **Leaves** usually less than 5 cm across, orbicular in outline, 5-lobed, cordate to truncate at base, with blunt teeth, hairy on both surfaces but often becoming glabrous with age; petioles puberulent and sparsely glandular with stalked glands. **Flowers** 1-2(-3) from nodal branchlets on old wood; peduncles short, puberulent; pedicels short, glabrous; bracts small, shorter than or equal to the pedicels; ovary glabrous; hypanthium glabrous, at least as long as wide; sepals narrowly diverging, white; petals white, shorter than the sepals; stamens about as long as the petals; style hairy below, divided above. **Fruit** glabrous, red to black. Plate 30, figure 3.

Range: Alaska to Newfoundland in the Boreal Forest Region; southward into the northeastern and north-central United States; across northern and central British Columbia and southward along the Rocky Mountains and adjacent ranges to southeastern British Columbia; recorded as far south as Yahk. Associated with moist woods and prairies at moderate elevations.

Ribes cereum Douglas **Waxy Currant, Squaw Currant**

Shrub 1-1.5(-2) m tall. Branches hairy and sticky, becoming glabrous with age, dark grey to brown. **Leaves** 12-25 mm across, fan-shaped, subcordate or truncate at base, very shallowly 3-lobed; the lobes broadly rounded, crenate, greyish green, sparsely hairy and often glandular on both sides and on the petioles. **Racemes** shorter than the leaves, most with 2-4 flowers; bracts broad, longer than the pedicels, toothed in ours; ovary glandular or smooth; hypanthium tubular, pubescent, whitish; sepals very short, ciliate, recurved, white; petals minute, white, scarcely visible between the sepal bases; anthers with apical cuplike glands; styles connate to tips. **Fruit** red, shiny. Plate 30, figure 4.

Range: Southern British Columbia southward to California and eastward to Montana. In British Columbia, east of the Coast and Cascade ranges in the dry interior, northward to Kamloops and westward to Lytton, on dry, gravelly slopes and in grass and sagebrush steppe communities.

Ribes aureum Pursh   Golden Currant

Erect shrub, 1-3 m tall, with glabrous branches. **Leaves** distinctly 3-lobed, with open sinuses, cuneate to truncate at base; the lobes coarsely and bluntly toothed at their tips, or entire; glabrous, or ciliate and pubescent when young. **Racemes** from leafy lateral branchlets, 25-75 mm long, with 5-18 flowers; peduncle sometimes puberulent; bracts elliptic to lanceolate, usually longer than the pedicels. **Flowers** 15-16 mm long; ovary glabrous; hypanthium 6-8 mm long, yellow; sepals 5-6 mm long, yellow; petals purplish or yellow, less than half as long as the sepals; stamens as long as petals, the filaments shorter than the anthers; styles connate to the tip. **Fruit** glabrous, red, black or yellow. Plate 30, figure 5.

Range: Central Washington state east of the Cascade Range to California, southern Alberta and Saskatchewan and Montana. Introduced in British Columbia and occasionally found wild in dry, open places (Princeton, Saltspring Island).

Ribes bracteosum Douglas *ex* Hooker   Stink Currant
                                                            Blue Currant

Erect aromatic shrub, 1-3 m tall. **Leaves** orbicular, 10-20 cm across, cordate at base; 5- to 7-lobed, the lobes acute, the margins sharply doubly toothed; sparsely hairy and glandless above, with sessile resinous glands beneath. **Racemes** ascending to erect, 10-23 cm long, many-flowered. Lower bracts leaflike, where longer than the pedicels; upper bracts lanceolate and shorter. Ovary puberulent and with abundant sessile glands; hypanthium saucer-shaped, 1-1.5 mm deep, puberulent and sometimes sparsely glandular. Sepals pale green to whitish, spreading; petals minute, truncate, narrowed to base, white. Stamens equal to or slightly exceeding petals; style divided about halfway to base. **Fruit** blue-black, with a white bloom and numerous sessile yellowish glands, attached along a now-pendant raceme. Plate 31, figure 1.

Range: Alaska to California from the Cascade and Coast ranges to the coast; typically found along streams and in moist, shady alluvial forests.

Ribes hudsonianum	**Hudson Bay Currant**
Richardson in Franklin	**Northern Black Currant**

Aromatic shrub, 0.5-2 m tall, with ascending or decumbent branches; twigs pale greenish to olive, sprinkled with minute yellow sessile resinous glands; older stems pale greyish, with dark dots; the bark ultimately shredding. **Leaves** 4-10 cm wide, wider than long; 3-lobed or occasionally 5-lobed; cordate or rounded at base, acute at apex, sharply dentate or serrate, glabrous above, resinous-dotted and thinly hairy beneath. **Racemes** erect or ascending, 12- to 23-flowered. Bracts hairy, linear, usually about half (or less) as long as the pedicels; occasionally the lower bracts longer. **Flowers** 7-8 mm long; ovary glabrous, commonly resin-dotted, with minute sessile, shiny glands; hypanthium funnel-shaped, 1-1.5 mm long, resin-dotted, puberulent; sepals white, puberulent, spreading less than 90° from flower axis; petals spatulate, about half as long as the sepals, white; stamens and styles slightly shorter than the petals; styles united to the tips, or nearly so. **Fruit** black, glabrous, usually dotted with pale sessile resin-glands, without a bloom, 7-12 mm long. Plate 31, figure 2.

Two intergradient varieties are weakly distinguished as follows:

1a Leaves hairy above as well as beneath; ovary often glandless. Common in central and northern British Columbia and widespread in the Boreal Forest Region .... **var.** *hudsonianum*

1b Leaves glabrous above and hairy beneath only along the veins; ovary and fruit with sessile resinous glands. Commoner from southern British Columbia southward to California and Utah .... **var.** *petiolare* **(Douglas) Janczewski**

Range: Alaska to Hudson Bay in the Boreal Forest Region and southward in the west to California and Utah. Found in moist woods and along streams and lakeshores.

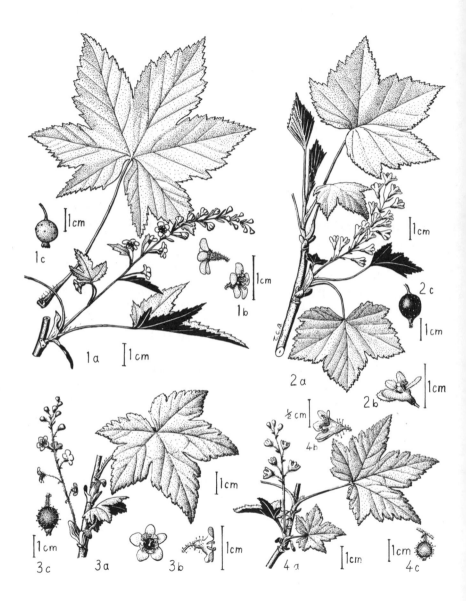

Plate 31 Currants. Figures: (1) *Ribes bracteosum;* (2) *R. hudsonianum;* (3) *R. laxiflorum;* (4) *R. glandulosum.* a, flowering branches; b, flowers; c, fruit.

Ribes laxiflorum **Pursh**   **Trailing Black Currant**

Shrub with ascending to spreading or trailing stems; twigs grey or brown, smooth. **Leaves** 2.5-12 cm across, orbicular in outline, deeply cordate at base; 5-lobed, the lobes acute at apex and coarsely serrate to biserrate; glabrous above, slightly hairy beneath and ciliate; petioles glandular-hairy. **Racemes** erect, 6- to 18-flowered, from lateral branchlets without leaves or with reduced leaves; bracts narrow, shorter than the pedicels. **Flowers** reddish; ovary with dark stalked glands and minute white puberulence; hypanthium saucer-shaped, 1 mm deep; sepals ovate to orbicular, spreading at right angles to the flower axis, red, 2.5-3 mm long; petals widely triangular to crescent-shaped, reddish; stamens and styles equalling the petals, the styles connate about half their length. **Fruit** blue-black to dark purple, with scattered gland-tipped bristles and thin puberulence, often with a bloom. Plate 31, figure 3.

Range: Moist woods and stream banks. Alaska to California along the coast and eastward through British Columbia to western Alberta and Idaho.

Ribes glandulosum **Grauer**   **Skunk Currant**
                                                 Fetid Currant

Fetid shrub up to a metre tall, with unarmed, prostrate to ascending stems. **Leaves** 2.5-8 cm across, rather deeply 5-lobed; the lobes acute at apex, biserrate; leaf blade thin, thinly hairy to glabrescent on both sides; petiole glandular-hairy toward base. **Racemes** ascending to erect from lateral buds, with a few small leaves; 6- to 10-flowered; bracts shorter than the pedicels. **Flowers** small; pedicel and ovary with numerous stalked glands, otherwise glabrous or thinly puberulent; hypanthium glabrous, less than 1 mm deep; sepals diverging, longer than wide, white; petals spatulate to cuneate, longer than wide, white to occasionally red; stamens equalling the petals; styles divided almost to base, equalling the petals. **Fruit** red, with many stalked glands. Plate 31, figure 4.

Range: Boreal-transcontinental: Alaska to Newfoundland and southward to Minnesota. In British Columbia, widespread in the northern and central interior, and southward to the Horsefly Lake area. Found in moist to dry coniferous woods.

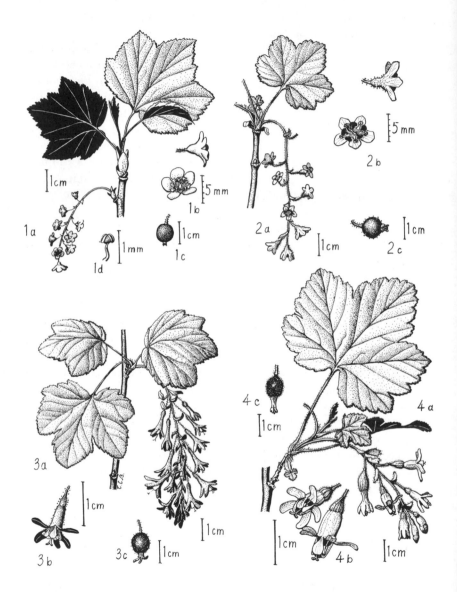

Plate 32 Currants. Figures: (1) *Ribes triste;* (2) *R. howellii;* (3) *R. san-guineum;* (4) *R. viscosissimum.* a, flowering branches; b, flowers; c, fruit; d, stamen.

Ribes triste Pallas   **American Red Currant**
Swamp Currant, Red Swamp Currant

Shrubs up to a metre tall, with ascending or reclining stems; bark grey, and shredding with age. Twigs greyish, smooth. **Leaves** 4-12 cm across, wider than long, cordate or truncate at base; distinctly but shallowly 3-lobed, rarely obscurely 5-lobed; acute to obtuse at apex, margins with blunt teeth, glabrous above, sparsely hairy on the veins beneath, sometimes with scattered sessile glands; petioles finely hairy. **Racemes** usually on leafless shoots from lateral buds, drooping, 6- to 15-flowered; the peduncle and pedicels finely puberulent and with scattered short-stipitate glands; bracts shorter than pedicels, glabrous or hairy. **Flowers** variously reddish to greenish; ovary glabrous; hypanthium saucer-shaped, about a millimetre deep, glabrous; sepals orbicular, spreading, greenish to reddish, 2-2.5 mm long and at least as wide; petals less than 1 mm long, less than half as long as the sepals, cuneate-deltoid, reddish; stamens about equalling the petals, with anthers widely separated at base but converging at apex; styles short, up to as long as the stamens, united from a quarter to full length. **Fruit** smooth, red, 7-10 mm in diameter. Plate 32, figure 1.

Range: Wet meadows and swamps from eastern Asia and Alaska to Newfoundland through the Boreal Forest Region and southward to Oregon in the west. Widespread in the interior of British Columbia, but uncommon except in the northern and central parts.

Ribes howellii Greene   **Mapleleaf Currant**
(*R. acerifolium* Howell, not Koch)

Spreading or erect shrub, up to a metre tall. **Leaves** wider than long, with 3-5 rather rounded lobes, serrate, glabrescent above, puberulent on the veins and gland-dotted beneath; petioles finely puberulent and sparingly glandular. **Racemes** drooping from the leaf axils, 6- to 15-flowered, hairy and glandular with short-stalked glands; bracts shorter than or same length as the pedicels. **Flowers** reddish, 7-8 mm wide; ovary with stalked glands; hypanthium puberulent with stalked glands, bowl-shaped, as wide as or wider than deep. Sepals reddish or paler and red-dotted, obovate. Petals obovate, semicircular, and truncate or cuneate-based; about a third the length of the sepals; dark red. Stamens about as long as the petals or slightly longer, with dilated filaments; style glabrous, divided toward the tip. **Fruit** blue-black, sometimes glaucous, with scattered stalked glands. Plate 32, figure 2.

Range: Southern British Columbia to Oregon in the Coast and Cascade ranges and in the Selkirk Mountains in northern Idaho; on alpine and subalpine slopes and ravines. In this province, northward to Garibaldi Park and eastward to Manning Provincial Park; look for it in the southern Selkirk Mountains.

Ribes sanguineum Pursh   Red-flowered Currant

Erect shrub up to 3 m tall; the branchlets finely hairy, reddish brown. **Leaves** 3-9 cm across, generally wider than long, cordate at base; 3- to 5-lobed, the lobes rounded in outline and biserrate; thinly hairy above and densely so beneath; petioles hairy and sparsely glandular. **Racemes** arching or drooping, 3-10 cm long, 10- to 25-flowered, puberulent and sparsely glandular with stalked glands; bracts oblanceolate to obovate, irregularly toothed, glandular, about as long as the pedicels, red. **Flowers** about 9-15 mm long, red. Ovary greenish, densely glandular with stalked glands; thinly hairy. Hypanthium tubular; it and the sepals red, thinly hairy and glandular externally. Petals spatulate to cuneate, about half as long as the sepals, white to red; stamens as long as petals; style about 2 mm longer than the stamens, undivided or divided only at the extreme tip. **Fruit** blue-black, with a whitish bloom and with stalked glands that may disappear from the ripe fruit. Plate 32, figure 3. Photo 36.

Range: Southern British Columbia to California, from the coast to the Cascade Range. In British Columbia, typically a coastal species in dry wooded sites; northward on the coast to Port Neville Inlet, but also found at New Denver and on the Arrow Lakes in the interior. Popular as a cultivated ornamental shrub.

Ribes viscosissimum Pursh   Sticky Currant

Shrub up to a metre tall; bark grey; twigs sparingly to densely velvety-hairy and glandular with stalked glands. **Leaves** 2.5-7.5 cm across, orbicular to kidney-shaped, cordate at base; 3- to 5-lobed, the lobes rounded with obtuse apices; biserrate; variably hairy and glandular on both surfaces; petioles densely glandular-hairy. **Racemes** spreading, arching or ascending from leaf axils, or terminal on lateral branchlets; 3- to 13-flowered, 5-8 cm long, glandular-hairy. Bracts about as long as the pedicels, obovate to oblanceolate, toothed toward the apex, green, glandular-hairy.

Flowers about 2 cm long, white; ovary greenish, densely glandular with stalked glands and hairy; hypanthium deeply cup-shaped, pale greenish to white, sparingly glandular and hairy, 5-6 mm long. Sepals spreading, white, ovate, 6 mm long; petals half as long as the sepals or slightly longer, obovate, white; stamens as long as the petals, the anthers with apical cuplike glands; style glabrous, equalling the petals and stamens or slightly longer, undivided almost to the tip. **Fruit** black, glandular-hairy, without a bloom in ours, 8-11 mm long. Plate 32, figure 4.

Range: British Columbia to California and eastward to southwestern Alberta, Montana and Colorado, from the Cascade Range to the Rocky Mountains; in British Columbia, northward to Lillooet. Found in subalpine coniferous forests, in both dry and wet sites.

Hydrangeaceae   Hydrangea Family

Shrubs with opposite simple leaves without stipules. Flowers in flat-topped panicles or racemes; the calyx cup adherent to the ovary but not extending beyond it as a hypanthium; sepals and petals 4-10; stamens usually numerous; ovary partly to completely inferior, of 3-10 carpels, with many or few ovules. Fruit a capsule with few to many seeds.

A family, formerly included in the Saxifragaceae, of a dozen genera and 75 species; distributed in subtropical and north temperate latitudes.

Philadelphus   Mock-orange, Syringa

Flowers in panicles or racemes terminating short lateral branches; showy and fragrant; hypanthium not developed; stamens numerous, free; ovary almost inferior; styles 4, connate about half their length, the free parts stigmatic on their inner sides.

Philadelphus lewisii Pursh   Mock-orange

Tall deciduous shrub, up to 5 m tall, with orange stems and white pith. **Leaves** short-petioled, ovate, entire or distantly dentate; with 5 strong veins from close to the base; thinly hairy or glabrous except for hairs in the angles of the veins beneath. **Flowers** in racemes terminating short lateral branches; pedicels glabrous, usually less than 1 cm long; sepals and petals 4 each; petals white, larger than the sepals; stamens 25-50; styles connate at least half their length. **Fruit** a capsule, half to two-thirds inferior, splitting down the locules. Plate 33, figure 1.

Range: From southern British Columbia and southwestern Alberta to Montana and California. Across southern British Columbia from Vancouver Island to the Rocky Mountains; northward to Savary Island on the coast and to Lillooet in the interior. Found in a variety of habitats, but preferring open or semi-open rocky ground. Photo 32.

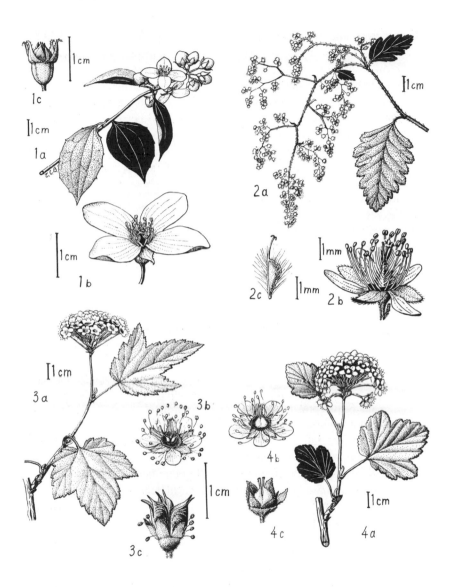

Plate 33 Figures: (1) *Philadelphus lewisii;* (2) *Holodiscus discolor;*
(3) *Physocarpus capitatus;* (4) *P. malvaceus.* a, flowering branch; b, flower;
c, fruit.

Rosaceae   Rose Family

Shrubs, trees and herbs, mostly deciduous, often thorny. Leaves mostly alternate, simple or compound; stipules usually conspicuous, rarely wanting. Flowers usually perfect, radially symmetrical; the bases of the sepals, petals and stamens united into a cuplike or saucerlike hypanthium that is commonly lined with a nectar-producing disc. The hypanthium bears the free parts of the appendages on its rim. Sepals and petals usually 5 each; stamens 5 to many. Ovary superior or inferior, of one to many free or united carpels. Fruits diverse: follicle, achene, drupe, pome or hip. The treatment of the Rose family here is based mainly on that of Taylor (58).

Key to the Genera of Rosaceae
1a Leaves simple.
  2a Fruit dry at maturity; follicles or achenes.
    3a Flowers in clusters; fruit splitting open except in *Holodiscus.*
      4a Shrub erect; leaves serrate or shallowly lobed.
        5a Leaves palmately veined; stipules present but deciduous
        . . . . *Physocarpus*
        5b Leaves pinnately veined; stipules absent.
          6a Leaves finely or sharply serrate, rounded at base;
          carpels glabrous or finely puberulent, forming follicles
          . . . . *Spiraea*
          6b Leaves shallowly lobed or biserrate, wedge-based;
          carpels stiffly long-hairy, forming achenes . . *Holodiscus*
      4b Shrub prostrate; leaves deeply palmately cleft . . . . . *Luetkea*
    3b Flowers solitary; fruit achenes.
      7a Erect, tall shrub with cuneate leaves; pedicels less than 1 cm
      long; fruit a single short-styled achene . . . . *Purshia*
      7b Prostrate dwarf shrub with elliptic to lanceolate leaves;
      pedicels 5-50 cm tall; fruit a cluster of achenes with long,
      plumose styles . . . . *Dryas*
  2b Fruit fleshy at maturity.
    8a Fruit a pome.
      9a Flowers in racemes . . . . *Amelanchier*
      9b Flowers in corymbs or umbels.
        10a Branches sharply spiny; fruit containing stony nutlets
        . . . . *Crataegus*
        10b Branches not spiny, although often with stiff blunt spurs
        ending in buds.

11a Leaves entire, 1-4 cm long; shrubs; fruit containing
nutlets . . . . *Cotoneaster*
11b Leaves serrate, 4-10 cm long; trees or big, coarse
shrubs; fruit containing paired seeds in cartilaginous
chambers.
    12a Fruit a pear, tapering to base, its flesh gritty; styles
    free; bark dark grey to blackish, rough, finally break-
    ing into small rectangular plates . . . . *Pyrus*
    12b Fruit an apple, indented or rounded at base, its
    flesh not gritty; styles united at base; bark brown,
    breaking into elongate strips or irregular plates
    . . . . *Malus*
8b Fruit a drupe or an aggregate of drupelets.
    13a Leaves entire; twigs with chambered pith . . . . . . *Oemleria*
    13b Leaves serrate; pith solid.
        14a Leaves lobed; drupelets many per flower . . . . *Rubus*
        14b Leaves unlobed; drupe 1 per flower . . . . *Prunus*
1b Leaves compound.
    15a Fruit dry, of exposed achenes . . . . *Potentilla*
    15b Fruit fleshy.
        16a Fruit a pome . . . . *Sorbus*
        16b Fruit an exposed aggregate of drupelets . . . . *Rubus*
        16c Fruit an aggregate of seedlike achenes enclosed in a fleshy
        receptacle cup (a "hip") . . . . *Rosa*

Physocarpus   Ninebarks

Shrubs with thin bark exfoliating in layers (whence the name *Ninebark*).
Leaves alternate, with small stipules; simple; palmately veined and lobed,
usually bearing branched, star-shaped hairs. Flowers terminal in dense,
broad umbel-like racemes on short lateral branches. Hypanthium shallow,
with a disc, bearing 5 sepals, 5 petals that are usually white, and 20-40
stamens; the centre of the disc is occupied by 1 to 5 separate superior
ovaries. In fruit the ovaries form inflated pods, each opening by splitting
along both sutures to release 2-5 slender, hard, shiny seeds.

Key to the Species of *Physocarpus*
1a Pods usually 2, white-hairy . . . . *P. malvaceus*
1b Pods 3-5, glabrous . . . . *P. capitatus*

Physocarpus capitatus **(Pursh) Kuntze**   **Pacific Ninebark**

Shrub up to 3 m tall, the lowest branches sometimes rooting. **Leaves** with 3-5 acute or obtuse lobes, with a biserrate margin, usually pubescent beneath. **Flowers** in many-flowered, dome-shaped racemes, with tomentose pedicels; the flowers about 1.2 cm across, the petals usually white. **Fruit** 3-5 small pods barely joined at base; acuminate, glabrous at maturity, longer than the sepals. Plate 33, figure 3. Photo 34.

Range: Southern British Columbia west of the Cascade Range and along the coast, including Vancouver Island, northward to Bella Coola, inland to Manning Provincial Park and southward along the coast to California. Also in the Columbia Forest Region, from Revelstoke, Shuswap Lake and the Arrow Lakes southward into Idaho. Commonly near streams or lakes, preferring moister habitats than *P. malvaceus.*

Physocarpus malvaceus **(Greene) Kuntze**   **Mallow Ninebark**
(P. pauciflorus **Piper)**

Shrub up to 2 m tall, with pubescent branchlets. **Leaves** orbicular to broadly ovate, cordate to truncate at base; shallowly 3- to 5-lobed, with broad, rounded, doubly crenate-dentate lobes; stellate-pubescent beneath and sometimes above; rarely glabrous. **Flowers** in dense, umbel-like racemes, white, 1-1.4 cm across. **Fruit** a pair of flat pods, often joined for half their length; no longer than the sepals; covered with white star-shaped hairs and with erect, glabrous beaks formed from the styles. Plate 33, figure 4.

Range: In the southern interior of British Columbia from the Okanagan Valley eastward across the South Kootenay region to the Rocky Mountains; also in southwestern Alberta, eastern and central Washington, Montana and Utah. Found in thickets and dry open slopes.

Holodiscus

Shrubs with alternate, simple leaves, without stipules. Leaves pinnately veined and toothed or lobed. Flowers small, very many in showy whitish panicles; calyx deeply 5-lobed; petals 5; stamens 20; carpels 5, separate, hairy, their styles about 1 mm long, with forked stigmas. Fruit stiffly hairy achenes.

Holodiscus discolor (Pursh) Maximowicz   Ocean-spray

Shrub 1-6 m tall with grey bark; branches arching, light brown, slightly hairy. **Leaves** 4-7 cm long by 2-5 cm wide; ovate, cuneate at base, obtuse or rounded at apex; shallowly pinnately lobed, the lobes toothed apically; glabrous above, slightly hairy beneath. **Flowers** very many, creamy white, 3-5 mm wide, in large pendulous, open panicles. **Fruit** shortly stiped achenes with long, stiff hairs projecting from the margins. Plate 33, figure 2. Photo 33.

Range: Across southern British Columbia south of lat. 51°N, including Vancouver Island, eastward to Montana and southward to California. Prefers lightly wooded rocky ground.

Spiraea   Spiraea

Shrubs of varying habit; ours with ascending thin, wiry branches. Leaves alternate, simple, without stipules, pinnately veined and serrate. Flowers small, in compact terminal panicles; calyx 5-lobed, persistent; petals 5; stamens numerous; carpels 5. Fruit small follicles, opening along the ventral suture, containing few to several seeds.

Key to the Species of *Spiraea*
1a Inflorescence cylindrical to conical.
   2a Inflorescence narrowly spirelike, dense; flowers deep pink to reddish . . . . *S. douglasii*
   2b Inflorescence conical, about as wide as tall, relatively loose; flowers pale pink to whitish . . . . *S. pyramidata*
1b Inflorescence flat-topped or broadly rounded.
   3a Flowers red. Inflorescence broadly rounded to dome-shaped
   . . . . *S. densiflora*
   3b Flowers white.
      4a Peduncles and hypanthium finely pubescent. Sepals reflexed. Inflorescence dome-shaped. Northwestern British Columbia
      . . . . *S. beauverdiana*
      4b Peduncles and hypanthium glabrous. Sepals ascending. Inflorescence flat-topped. Central interior of British Columbia and southward and eastward . . . . *S. betulifolia*

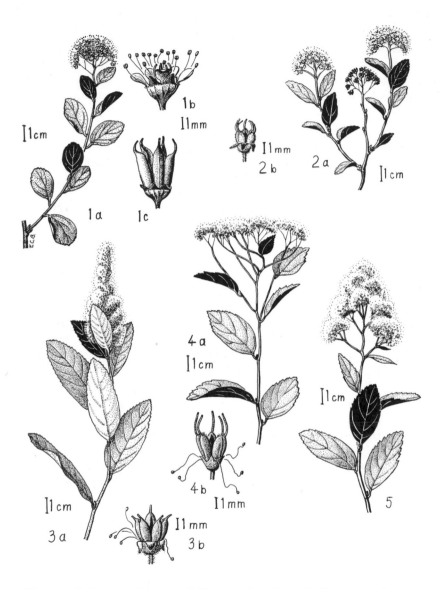

Plate 34 Spiraeas. Figures: (1) *Spiraea densiflora;* (2) *S. beauverdiana;* (3) *S. douglasii;* (4) *S. betulifolia;* (5) *S. pyramidata:* a, flowering branches. 1b, flower; 1c, 2b, 3b, 4b, fruits.

Spiraea densiflora **Nuttall** **Mountain Spiraea**
ex **Torrey & Gray**

Shrub 30-90 cm tall, with glabrous slender branches. **Leaves** 1.5-3 cm long, oval; round or obtuse at apex, round or cuneate at base; serrate near the apex, with rounded teeth; glabrous on both surfaces. **Flowers** red, 4-5 mm wide, numerous in round, dense inflorescences; hypanthium glabrous outside, puberulent within; sepals erect to spreading, not reflexed, usually glabrous; petals obovate, 1.5-2 mm long; stamens about 3 mm long. **Fruit** 5 glabrous shiny follicles, 3-4 mm long, exserted beyond the sepals. Plate 34, figure 1. Photo 35.

Range: Open slopes at subalpine altitudes, across southern British Columbia from Vancouver Island to the Rocky Mountains and northward to Revelstoke; also southward to California and eastward to Montana.

Spiraea beauverdiana **Schneider** **Beauverd Spiraea**
(*S. stevenii* **[Schneider] Rydberg)**

Low, branched shrub, usually under 30 cm tall, with incurved pubescence on young branches. **Leaves** 1.5-4 cm long, elliptic, rounded at ends; shallowly serrate at apex or nearly full length; glabrous above, glabrous or pubescent beneath. **Flowers** 5 mm wide, white; hypanthium puberulent; sepals pubescent dorsally, becoming reflexed; petals orbicular, longer than sepals; stamens 10-20, the filaments twice or more as long as the petals. **Fruit** puberulent follicles about 2 mm long. Plate 34, figure 2.

Range: In alpine or muskeg habitats from Alaska to the Northwest Territories and extreme northwestern British Columbia. Found at timberline and above on the Haines Highway.

Spiraea betulifolia **Pallas** **Flat-topped Spiraea**
Var. *lucida* **(Douglas** *ex* **Hooker) C.L. Hitchcock**
(*S. lucida* **Douglas** *ex* **Hooker)**

A low shrub, usually less than 60 cm tall. **Leaves** 1-8 cm long, ovate or oval; rounded to acute at apex; serrate above mid-length, or sometimes entire; green above, paler beneath. **Flowers** about 5 mm wide, in a flat compound corymb 5-10 cm wide; hypanthium glabrous; sepals broadly triangular, erect to spreading, rarely reflexed; petals circular, white, about 1.5 mm long; stamens 20, up to 5 mm long. **Fruit** glabrous, shiny, beaked follicles about 3 mm long, with styles 0.5-2 mm long. Plate 34, figure 4. Photo 39.

Range: Typically in dry, open woods at moderate elevations in the Coast and Cascade ranges and through the interior of British Columbia northward to the Peace River valley, and at Agassiz and Bella Coola, west of the coastal ranges; also eastward to Saskatchewan and southward to Oregon and Wyoming. The typical variety, *betulifolia,* occurs in Asia.

Spiraea douglasii Hooker   Hardhack

Shrub 1-2.5 m tall, spreading by rhizomes to form extensive colonies. **Leaves** 3-8 cm long, narrowly elliptic; serrate around the apical half, rounded at base and apex; glabrous above, densely puberulent to nearly glabrous beneath. **Flowers** about 4 mm wide, deep pink in dense narrowly conical or cylindrical panicles, 2-5 times as long as wide; hypanthium tomentose to nearly glabrous; sepals reflexed; petals obovate. **Fruit** 5 glabrous, shiny, short-beaked follicles, 2-3 mm long. Plate 34, figure 3. Photo 38.

Range: In moist to swampy ground at low to moderate altitudes; from southern Alaska to California along the coast and in the interior of British Columbia as far as the West Kootenay region and northward to lat. 56°N.
There are two varieties in British Columbia. Var. *douglasii,* with leaves greyish-downy beneath, is the common variety from the coast to the Coast and Cascade ranges. Var. *menziesii* (Hooker) Presl, with leaves glabrous or nearly so beneath, is the common variety east of the Coast Mountains, often in Sphagnum bogs.

Spiraea pyramidata Greene   Pyramid Spiraea

Shrub 30-100 cm tall. **Leaves** 4-9 cm long, elliptic, serrate toward the apex, slightly hairy beneath. **Flowers** pale rose to nearly white; borne in a conical panicle about as wide as tall; flowers about 4 mm wide; hypanthium thinly pubescent; sepals about as long as the hypanthium, reflexed;

petals twice as long as the sepals, orbicular; ovary usually finely puberulent, sometimes glabrous. **Fruit** not seen in British Columbia. Plate 34, figure 5.

Range: In dry to moist sites in the southern interior of British Columbia, southward to Oregon and eastward to Idaho. Widespread but not abundant in the southern interior of British Columbia, northward to around Hazelton (at lat. 55°N).

Note: *S. pyramidata* has been suspected of being of hybrid origin, between *S. douglasii* and *S. betulifolia.*

Luetkea

Low subshrub, growing in mats, with long, creeping stems and 3-cleft leaves. Flowers small, white, in erect racemes. Sepals 5, oval; stamens 20, sometimes united at their bases, shorter than the petals, arising under the edge of the prominent disc; pistils 5, as long as the stamens. Fruit of 5 follicles, opening along the ventral suture and for a short distance down the dorsal suture.

Luetkea pectinata (Pursh) Kuntze — Partridgefoot / Meadow Spiraea

The only species. Matted, with long, slender creeping woody stems. **Leaves** without stipules, glabrous; cleft into narrow lobes above a petiole equally wide. **Flowers** 6-8 mm across, in erect terminal racemes 2-5 cm long, on erect basally leafy shoots 5-20 cm tall. **Fruit** follicles about 5 mm long, with 4 seeds. Plate 35, figure 1. Photo 40.

Range: Alaska to California and eastward to Montana. Almost throughout British Columbia at alpine elevations; abundant and constant in damp alpine meadows.

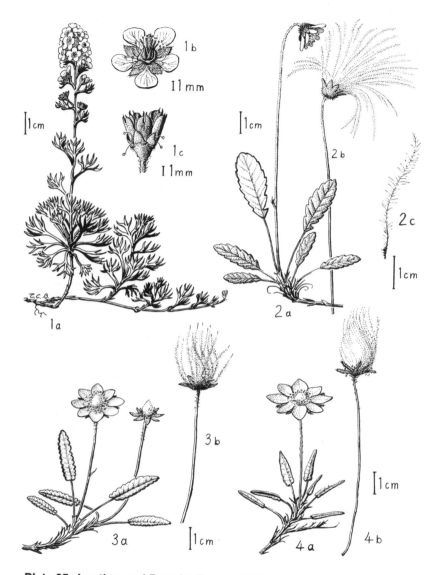

Plate 35 *Luetkea* and Dryads. Figures: (1) *Luetkea pectinata:* 1a, flowering branch; b, flower; 1c, fruit (follicles); (2) *Dryas drummondii:* 2a, flowering shoot; 2b, head of achenes; c, achene; (3) *D. octopetala:* 3a, flowering shoot; 3b, head of achenes; (4) *D. integrifolia:* 4a, flowering shoot; 4b, head of achenes.

Dryas   Dryads, Mountain-avens

Prostrate dwarf, mat-forming shrub. Leaves basal, alternate, leathery, simple, evergreen, usually glabrous above and tomentose beneath. Stipules present. Flowers solitary on erect pedicels; sepals and petals 8-10 each; stamens and carpels numerous. Fruit a head of achenes with long, plumose stylar appendages.

Key to the Species of *Dryas*
1a Leaf tapered to petiole, with crenate margin; flower yellow
. . . . *D. drummondii*
1b Leaf cordate or truncate at base; flower white.
  2a Leaf margin distinctly crenate full length; midvein beneath with
    brown hairs or glands . . . . *D. octopetala*
  2b Leaf margin entire or nearly so; midvein beneath without brown
    hairs or glands . . . . *D. integrifolia*

Dryas drummondii   **Yellow Mountain-avens**
Richardson in Hooker   **Yellow Dryad**

Leaves elliptic, tapered to base, rounded at apex, with crenate margin; glabrous and wrinkled above, white-tomentose beneath and glandless beneath. Pedicel 15-20 cm tall, with a few small bracts; taller in fruit. **Flowers** nodding, with ovate sepals and yellow petals, never fully opening. Plate 35, figure 2. Photo 41.

Range: Alaska and Northwest Territories to Quebec and southward to Oregon and Montana. Widespread in British Columbia; commonly on gravel bars in rivers and stony alluvial flats and on glacial moraines, often at low elevations.

Dryas integrifolia **Vahl**   **Entire-leaved Mountain-avens**

Leaves lanceolate, broadest near the cordate or truncate base; acute at apex, entire or shallowly crenate near base; white-tomentose and glandless beneath. Pedicel 2-10(-15) cm tall, bractless and glandless. **Flowers** erect, with lanceolate sepals and showy white, ovate petals, opening wide. Plate 35, figure 4.

Range: Alaska to Greenland in the arctic tundra. In British Columbia, generally north of lat. 55°N, but southward in the Rocky Mountains to the Montana border. Found in alpine tundra, especially where disturbed.

Dryas octopetala L.	**White Mountain-avens** **White Dryad**

Leaves ovate-oblong, cordate to truncate at base, rounded at apex, distinctly crenate all along the margin; wrinkled above, white-tomentose beneath, with scattered brown hairs or glands on the midrib. Pedicel 3-15 cm tall, bractless, with black glandular hairs near the top. **Flowers** erect, with lanceolate sepals and 8 white petals, showy, opening wide. Plate 35, figure 3. Photo 37.

Range: Extremely variable, with many regional forms. Circumboreal, and southward in the mountains to Oregon and Colorado. Widespread in British Columbia in rocky alpine tundra.

Potentilla	**Cinquefoils, Potentillas**

A genus of mainly low, herbaceous species, but with a few shrubby members. Leaves usually compound, with stipules. Flowers solitary or in bracted cymose panicles. Hypanthium bearing 5 bractlets that alternate with the sepals, 5 petals and numerous stamens. Carpels many, 1-ovuled, on a central projection of the axis of the flower. Fruit a head of 1-seeded achenes.

Key to the Shrubby Species of *Potentilla*
1a Terrestrial, abundantly branched shrub; normally erect, but sometimes prostrate in alpine habitats. Leaflets entire. Petals yellow, longer than the sepals . . . . *P. fruticosa*
1b Amphibious to aquatic, sparingly branched shrub; trailing in water or over adjacent bushes. Leaflets serrate. Petals dark purple, shorter than the sepals . . . . *P. palustris*

Potentilla fruticosa L.   Shrubby Cinquefoil

Erect, dark-stemmed deciduous shrub, 30-100 cm tall, or prostrate in high alpine habitats. **Leaves** pinnately compound; leaflets (3-)5(-7), narrowly elliptic, entire, greyish green and hairy, especially beneath. **Flowers** scattered or in small clusters, yellow, 20-25 mm wide; sepals triangular, shorter than the narrow bractlets and shorter than the petals. **Fruit** a head of achenes with long, straight, stiff hairs. Plate 36, figure 1. Photo 42.

Range: Circumboreal, and southward in the mountains to California and New Mexico. Almost throughout British Columbia, from moderate to alpine elevations; adaptable to different habitats, but commonly in open damp areas.

Potentilla palustris (L.) Scopoli   Marsh Cinquefoil

Prostrate amphibious or aquatic shrub with long, sparingly branched stems trailing in water and over adjacent vegetation. **Leaves** with usually 7 leaflets that are oblanceolate to obovate, bluntly tipped, serrate, pale green, glabrous or hairy; the bases of the petioles and stipules sheathing the stem. **Flowers** grouped in erect leafy-bracted flowering branches; sepals lanceolate, greenish or purple-tinged, 7-12 mm long; petals dark purple, shorter than the sepals. **Fruit** a head of glabrous, smooth achenes. Plate 36, figure 2.

Range: Circumboreal, and southward to California. Throughout British Columbia, in marshy habitats and along the margins of lakes and streams.

Purshia   Bitterbrush, Antelope Brush

Deciduous shrubs with radiating dark stems and small, scaly buds. Leaves alternate, partly fascicled; apparently simple but actually compound, with a single deciduous leaflet on a persistent petiole; cuneate and apically 3-lobed. Stipules minute, triangular, attached to the persistent petiole. Flowers perfect, solitary, terminating short lateral branchlets; hypanthium funnel-shaped; sepals 5, greenish; petals 5, spatulate, greenish yellow; stamens about 25; pistil usually 1, sometimes 2. Style persistent. Fruit a pubescent, spindle-shaped achene, sitting in the calyx cup; longer than the persistent sepals.

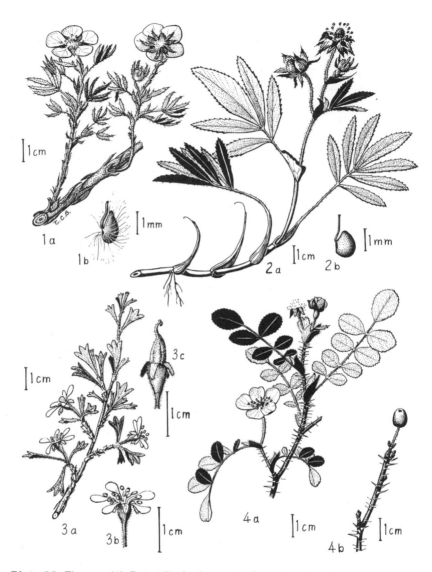

Plate 36 Figures: (1) *Potentilla fruticosa:* 1a, flowering branch; 1b, achene; (2) *Potentilla palustris:* 2a, flowering branch; 2b, achene; (3) *Purshia tridentata:* 3a, flowering branch; 3b, flower; 3c, fruit; (4) *Rosa gymnocarpa:* 4a, flowering branch; 4b, twig with fruit (hip).

186 – ROSACEAE

Purshia tridentata (Pursh) DC.   **Bitterbrush**
                                           Antelope Brush

Dark-stemmed, deciduous shrub up to 2.5 m tall; young branches pubescent. **Leaves** with very short (1-3 mm) persistent petioles that do not extend beyond the union with the stipules. The deciduous leaf blade 1-2 cm long, cuneate; apically 3-lobed, the lobes obtuse, the margin revolute; greyish to olive green and slightly pubescent above, whitish-tomentose beneath. **Flowers** 8-15 mm across; calyx villous and slightly glandular; petals obovate to spatulate, yellow; stamens 25. **Fruit** about 1 cm long. Plate 36, figure 3.

Range: Southern British Columbia to California, Colorado and New Mexico. In British Columbia, in arid sections of the Okanagan Valley from Kelowna southward and in the Rocky Mountain Trench south of lat. 50.5°N; under open Ponderosa Pine stands or dominating steppe vegetation on sandy to stony soils (Photo 50).

Note: Because of the similarity of their leaf forms, Bitterbrush can be confused with Big Sagebrush (*Artemisia tridentata:* see page 337 and Plate 75). The colour of the foliage and the bush provides the clearest answer to this question. Bitterbrush is a dull olive-green in colour, while Big Sagebrush is pale grey. In winter, Bitterbrush is leafless and presents a spectacle of radiating blackish stems and branches, while Big Sagebrush is evergreen and as grey in winter as in summer. They prefer different soil types and seldom grow together.

Rosa   **Roses**

Shrubs, usually with prickly stems; the prickles often differentiated into paired, relatively large, infrastipular prickles and finer prickles scattered on the internodes. Leaves alternate, odd-pinnately compound, stipulate, the stipules joined to the petioles. Flowers showy, solitary or in clusters on bracted peduncles. Hypanthium urn-shaped, contracted at the top and on its margin bearing the 5 sepals, 5 petals, normally pink on ours, and numerous stamens. Carpels many within the calyx tube; in fruit becoming achenes enclosed in the swollen fleshy calyx tube (the hip).

Key to the Species of *Rosa*

1a Infrastipular prickles not, or barely, distinct from other prickles. Stems often bristly.

  2a Sepals persistent on ripe fruit, attenuate to terminal flat appendages; petals 2-3 cm long; leaflets elliptic to obovate, more or less acute at ends . . . . *R. acicularis*

  2b Sepals shed from the ripening fruit, acute, unappendaged; petals 1-1.5 cm long; leaflets broadly elliptic to nearly orbicular, rounded at ends . . . . *R. gymnocarpa*

1b Infrastipular prickles present and clearly distinct from others.

  3a Infrastipular prickles normally placed, straight or nearly so. Stems reddish. Foliage not or sparingly glandular, not aromatic. Sepals not pinnately divided. Pedicels not bristly-glandular.

    4a Infrastipular prickles stout; commonly triangular in outline and about as wide-based as long. Flowers solitary on branchlets. Petals 2.5-4 cm long . . . . *R. nutkana*

    4b Infrastipular prickles slender, longer than wide at base. Flowers clustered. Petals 1-2.5 cm long.

      5a Sepals glandular, with gland-tipped bristles on back. From Cascade and Coast ranges to Vancouver Island . . . *R. pisocarpa*

      5b Sepals nearly always without glands on back. From Cascade and Coast ranges eastward . . . . *R. woodsii*

  3b Infrastipular prickles displaced downward on the internode, often single, stout, recurved to hooked. Stems greenish. Sepals often pinnately divided.

    6a Foliage strongly glandular, emitting a fruity aroma in warm weather. Pedicels bristly-glandular. Leaflet margin biserrate . . . . *R. eglanteria*

    6b Foliage not glandular or aromatic. Pedicels not bristly. Leaflet margin simply serrate . . . . *R. canina*

Rosa gymnocarpa **Nuttall**   **Baldhip Rose, Dwarf Rose**

Shrub with slender, bristly stems 0.6-1.6 m tall; the prickles straight, weak, the infrastipular ones barely distinguishable from the others. **Leaves** with 5-9 leaflets, the petioles glandular and often minutely prickly; stipules glandular-hairy on the margins. **Leaflets** 4-20 mm long, broadly elliptic; biserrate with fine gland-tipped teeth; glabrous. **Flowers** 20-35 mm across, solitary or in small clusters, deep rose-coloured; sepals triangular, without appendages, acute, deciduous after flowering. **Fruit** small, ovoid to almost spherical, berrylike, with no calyx when ripe. Plate 36, figure 4.

Range: Southern British Columbia to California and Montana. Widely distributed across southern British Columbia south of lat. 52°N, including Vancouver Island. Found in dry, open coniferous woods.

Rosa woodsii Lindley   Prairie Rose

Shrub up to 2 m tall, variably armed with prickles; internodal prickles, when present, are finer than the infrastipular ones, straight or curved. **Leaflets** 5-9, obovate, acute at base, acute to rounded at apex, simply serrate or biserrate. **Flowers** 1-5 in a terminal cluster, 30-45 mm across; pedicel and hypanthium glabrous; sepals 9-15 mm long, usually with a flat terminal appendage, glandular on the back. **Fruit** small with a short neck, crowned by erect or spreading sepals. Plate 37, figure 1.

Range: Interior of British Columbia and northeastward to the Mackenzie District, Northwest Territories, eastward to Saskatchewan and Wisconsin and southward to California and Texas. In British Columbia, northward to the Stikine River.

Rosa pisocarpa Gray   Swamp Rose, Clustered Wild Rose

Shrub 1-2 m tall, the stems armed with straight, slender, infrastipular prickles, or sometimes unarmed. **Leaflets** 5-7, 1-8 cm long, oblong to ovate, simply serrate; glabrous above, hairy beneath; without marginal glands; petioles short, hairy; stipules slightly glandular-toothed, hairy. **Flowers** 20-30 mm across, usually in terminal clusters; pedicels glabrous, slender; hypanthium glabrous; sepals lanceolate with flat terminal appendages, sometimes glandular, white-hairy within. **Fruit** small, subglobose, dark red, with short neck, 7-9 mm wide. Plate 37, figure 2.

Range: Southern British Columbia to California west of the Coast and Cascade ranges in swamps and moist thickets. In British Columbia, on Vancouver Island and the adjacent coast and in the Fraser River valley up to Yale.

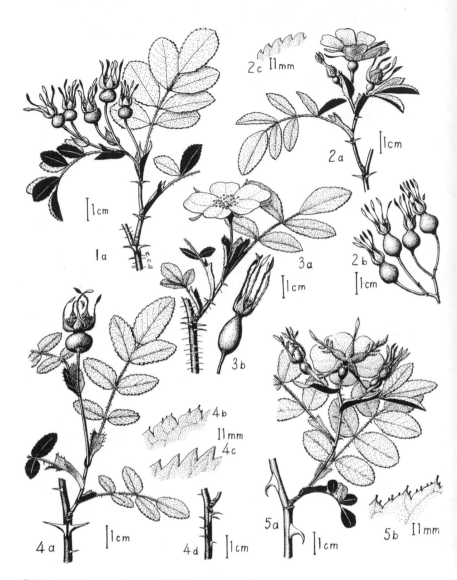

Plate 37 Roses. Figures: (1) *Rosa woodsii:* 1a, fruiting branch;
(2) *R. pisocarpa:* 2a, flowering branch; 2b, fruits; 2c, leaflet margin;
(3) *R. acicularis:* 3a, flowering branch; 3b, fruit; (4) *R. nutkana:* 4a, fruiting
branch; 4b, leaflet margin of var. *nutkana;* 4c, leaflet margin of var. *hispida;*
4d, stem and prickles of var. *hispida;* (5) *R. eglanteria:* 5a, flowering
branch; 5b, leaflet margin.

Note: In the absence of flowers or fruit, *R. pisocarpa* can usually be distinguished from the coastal variety of *R. nutkana* var. *nutkana*, with which it may grow, by its simply serrate leaflet margins; those of the *R. nutkana* population here nearly always display biserrate margins.

Rosa acicularis Lindley   Bristly Rose, Prickly Rose

Shrub 30-120 cm tall. Stems generally densely armed with straight slender prickles and bristles; the infrastipular prickles are commonly indistinguishable from the internodal ones and may be absent. **Leaves** with glandular-hairy petioles and rachises; leaflets 3-9, elliptic, acute at both ends or rounded at base, simply serrate or biserrate, hairy beneath; stipules often dilated and glandular-toothed. **Flowers** usually solitary, occasionally in twos or threes, 4-5 cm across; pedicels glabrous in ours; sepals entire or slightly lobed, up to 3 cm long, including the terminal appendage. **Fruit** usually ovoid or pear-shaped, with a pronounced neck, up to 2 cm long by 1 cm wide, crowned by the erect sepals. Plate 37, figure 3.

Range: Circumboreal, and southward along the Rocky Mountains to Colorado and New Mexico. Widespread and common across British Columbia east of the Coast Range.

Rosa nutkana Presl   Nootka Rose

Shrub 0.3-3 m tall, armed with stout, straight or curved prickles, the infrastipular prickles distinctly larger than the internodal ones. **Leaves** with 5-9 leaflets; stipules dilated, glandular; leaf rachis glandular, often prickly; leaflets oval, rounded at both ends or acute at tips, hairy and gland-dotted beneath, serrate. **Flowers** solitary or in twos or threes, 50-75 mm across; pedicel usually slightly glandular and bristly, hypanthium glabrous; sepals glandular-bristly, sometimes leaflike. **Fruit** usually globular, 1 cm or more in diameter. Plate 37, figure 4.

Key to the Varieties of *R. nutkana*
1a Infrastipular spines very stout. Leaflet margin doubly glandular- serrate. Coastal . . . . **var.** *nutkana*
1b Infrastipular spines rather slender. Leaflet margin simply serrate, at least distally, without glands. Widespread in the interior
. . . . **var.** *hispida* **Fernald**

Range: Common in thickets. Var. *nutkana* coastal from Alaska to California. Var. *hispida* across British Columbia east of the Coast range at lat. 56°N and southward to Colorado.

Rosa eglanteria L. (*R. rubiginosa* L.)   Sweetbrier, Eglantine

Coarse shrub; the stems stout and greenish, armed with scattered strong, hooked, internodally placed, often unpaired, infrastipular spines and sometimes with short, straight, internodal prickles. **Leaf** stalk and rachis finely prickly and densely glandular. Leaflets broadly elliptic, rounded to obtuse at base and apex, often less than 25 mm long; biserrate; glandular beneath and on margin; aromatic, the bush emitting a fruity aroma in warm weather. **Flowers** pink, 30-40 mm across; the short pedicel and sometimes the hypanthium with short, stiff, gland-tipped bristles; 2 or 3 of the sepals with minute leaflets. **Fruit** ovoid or subglobose; with sepals spreading or reflexed, often deciduous from the ripe fruit. Plate 37, figure 5.

Range: Introduced from Europe; established on the southern coast and in the Fraser River valley up to Hope.

Rosa canina L.   Dog Rose

Shrub resembling *R. eglanteria,* but not glandular and usually glabrous or nearly so in ours. Infrastipular spines hooked or curved. **Leaflets** more acute than in *R. eglanteria,* the margins usually simply and sharply serrate, without glands or sometimes with minute glands. **Sepals** reflexed over the unripe hip, dropping from the ripe hip. Styles of the achenes separate, protruding from a flattened disc. Plate 38, figure 1.

Range: Introduced from Europe; recorded on southeastern Vancouver Island and the Gulf Islands.

Rosa stylosa Desvaux

Practically indistinguishable from *R. canina,* differing in that the styles are coherent into a column that protrudes from a conical disc. Not illustrated.

Range: Introduced from Europe; recorded from Saltspring Island.

Note: *R. stylosa* and *R. canina* are so alike that when more living material is examined, some of British Columbian specimens thought to be *R. canina* may be found to be *R. stylosa*, or vice versa.

Rubus   Blackberries and Raspberries

Shrubs or herbaccous plants, often armed with prickles on stems and leaves. Leaves alternate, stipulate, simple or compound. Flowers usually perfect, axillary or in terminal clusters, mostly white; calyx usually 5-parted; petals usually 5; stamens many; carpels 5 to many on a central raised receptacle. Fruits are aggregates of small drupelets that may be shed from the floral receptacle, as in raspberries, or shed attached to the receptacle, as in blackberries.

Key to the Species of *Rubus*
1a Stems unarmed.
   2a Leaves normally simple.
      3a Stems erect; shrub to 1.5 m tall; leaves 10 cm or more across, 5-lobed .... *R. parviflorus*
      3b Stems trailing; low shrub under 0.5 m tall; leaves smaller, 3-lobed .... *R. lasiococcus*
   2b Leaves trifoliolate; main stems low-arching or prostrate, with ascending flowering branches .... *R. pubescens*
1b Stems armed with prickles.
   4a Leaves mostly simple; stems trailing .... *R. nivalis*
   4b Leaves normally compound.
      5a Stems usually erect, sometimes arching; fruit a raspberry: an aggregate of loosely united drupelets separating from the receptacle.
         6a Flowers solitary, petals pink to red; stems erect, orange, weakly prickly .... *R. spectabilis*
         6b Flowers clustered; petals white.
            7a Stems, petioles and inflorescence sticky with a dense covering of red, gland-tipped hairs and bristles; petals pink .... *R. phoenicolasius*
            7b Stems thinly short-hairy or glabrous, not sticky; petals white.

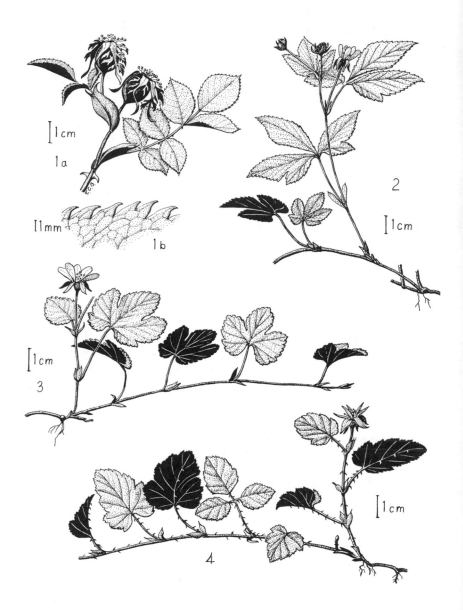

Plate 38 Figures: (1) *Rosa canina:* 1a, branch with subripe fruit; 1b, leaflet margin; (2) *Rubus pubescens;* (3) *Rubus lasiococcus;* (4) *Rubus nivalis.*

194 – ROSACEAE

8a Inflorescence corymbose; stems strongly pruinose,
with hooked prickles, sometimes arching; fruit black or
nearly so . . . . *R. leucodermis*
8b Inflorescence racemose; stems dark, scarcely glau-
cous, bristly, erect; fruit red . . . . *R. idaeus*
5b Stems arching or prostrate; fruit a blackberry: an aggregate of
separate drupelets adhering to the receptacle.
9a Stems trailing, with fine, straight prickles; leaves mostly trifoliolate, sometimes simple and 3-lobed . . . . *R. ursinus*
9b Stems arching, with strong, hooked prickles; leaves usually
palmately 5-foliolate.
10a Leaflets deeply, jaggedly dissected . . . . . . *R. laciniatus*
10b Leaflets not deeply dissected, broadly ovate to orbicular
. . . . *R. discolor*

Rubus pubescens Rafinesque   Trailing Raspberry

Low, thinly pubescent subshrub, with trailing and rooting, unarmed main
stems and ascending flowering branches. **Leaves** mostly trifoliolate, their
terminal leaflets narrowly rhombic; the laterals obliquely ovate, simply
serrate or biserrate, pubescent especially on the petioles. **Flowers** 1 to a
few, terminal on the erect branches; the sepals about 5-6 mm long,
reflexed; petals narrowly obovate, erect, up to 10 mm long, white; stamens many, short; carpels 20 or more. **Fruit** deep red, juicy, eventually
separable from the receptacle. Plate 38, figure 2.

Range: In moist woods from the central and northern interior of British
Columbia, north of Prince George, eastward through the Boreal Forest
Region to Newfoundland and southward along the Rocky Mountains to
Colorado.

Rubus lasiococcus Gray

Creeping subshrub with prostrate, slender, finely pubescent stems and
ascending, short flowering branches. **Leaves** simple, 3-lobed; the lobes
broad and rounded in outline, biserrate, finely pubescent on the petioles
and blades. **Flowers** 1 or 2 terminating the erect branches; the sepals
reflexed, 5-10 mm long; petals obovate, slightly longer than the sepals,
white; carpels 5-15, finely and densely white-puberulent. **Fruit** red,
puberulent, up to 1 cm across. Plate 38, figure 3.

Range: Woods in the mountains, from southern British Columbia to California in and west of the Cascade Mountains. In British Columbia, known from Vancouver Island and in Manning Provincial Park in the Cascade Mountains.

Rubus nivalis Douglas *ex* Hooker — Snow Bramble

Trailing subshrub with short, erect flowering branches; the stems, petioles and undersides of the leaf veins prickly with short hooked prickles. **Leaves** simple or occasionally trifoliolate, thinly pubescent, biserrate, the veins rather conspicuous. **Flowers** 1 or 2 terminating the short, erect flowering branches; sepals 7-9 mm long, ovate, reflexed, often tinged purplish; petals about as long as the sepals or slightly longer, narrowly lanceolate, dull pink to purplish, inconspicuous; stamens 15; carpels 4-9, puberulent. **Fruit** of a few relatively large drupelets. Plate 38, figure 4.

Range: In moist woods from British Columbia to Oregon, and in Idaho. In British Columbia, known on Vancouver Island and around Shuswap Lake and the Arrow Lakes.

Rubus parviflorus Nuttall — Thimbleberry

Shrub 1-2 m tall, forming colonies by suckering; the stems erect, unarmed, with shreddy bark. Branches glandular-hairy, the bud scales persistent around their bases. **Leaves** simple, palmately 5-lobed, 10-20 cm across, cordate at base, the margin biserrate, variably hairy on both sides, deciduous; petioles with red gland-tipped hairs; stipules lanceolate, 6-13 mm long. **Flowers** in small glandular-hairy corymbs, white, 4-5 cm wide. Sepals ovate, densely hairy and glandular, with linear terminal appendages; almost as long as the petals, which are ovate and up to 2.5 cm long. **Fruit** a dull red, convex, finely velvety-hairy raspberry, separating from the receptacle, edible but sour. Plate 39, figure 1. Photo 43.

Range: Alaska to California and eastward to the Dakotas and Lake Superior. Across British Columbia, from the coast to the Rocky Mountains and northward at least to the Peace River; from lowlands to subalpine elevations.

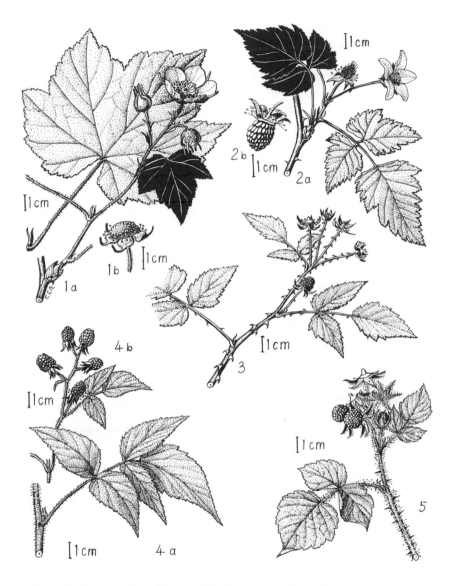

Plate 39 Raspberries. Figures: (1) *Rubus parviflorus:* 1a, flowering branch and leaf; 1b, fruit; (2) *R. spectabilis:* 2a, flowering branch; 2b, fruit; (3) *R. leucodermis;* (4) *R. idaeus:* 4a, leaf of first-year cane; 4b, fruit and leaf of fertile branch; (5) *R. phoenicolasius* in fruit.

Rubus spectabilis Pursh   Salmonberry

Shrub 1-5 m tall, with yellowish exfoliating bark; the stems weakly prickly, often unarmed on the upper branches. **Leaves** trifoliolate, deciduous; the lateral leaflets asymmetrical, shallowly lobed and biserrate, glabrous or sparsely hairy; stipules linear, 5-10 mm long. **Flowers** solitary or few, on short lateral branchlets with persistent bud scales around their bases; flowers about 25 mm across, nodding; sepals ovate, hairy; petals rosy red, ovate or elliptic. **Fruit** a dark purplish, red or yellow glabrous raspberry, edible and sweet. Plate 39, figure 2.

Range: Moist woods and stream banks along the coast from Alaska to California, mostly west of the Coast and Cascade ranges; occasionally penetrating farther east along the river valleys, as far as Smithers and Hope.

Rubus leucodermis Douglas *ex* Torrey & Gray   Blackcap \ Black Raspberry

Shrub 1-2 m tall, the ascending and arching branches very glaucous, armed with short, stiff, straight or somewhat recurved prickles and sometimes rooting at the tips. **Leaves** with 3-5 leaflets that are acute, rounded or cordate at base, lobed and biserrate, glabrous above, densely white-tomentose beneath, with prickles on petioles and midveins; stipules linear, 5-8 mm long. **Flowers** few in terminal corymbs and sometimes in the upper axils; sepals acuminate, reflexed, hairy; petals shorter, white. **Fruit** a purple to black raspberry; covered with a fine greyish down; separating from the receptacle, edible. Plate 39, figure 3.

Range: Dry, rocky soils in woods from Alaska to California along the coast and eastward to Montana; across southern British Columbia from the coast to the Selkirk Range.

Rubus idaeus L.   Raspberry

Erect shrub 1-2 m tall. The stems variably prickly with short prickles, often bristly; glandular, somewhat glaucous. **Leaves** on young, non-flowering canes pinnately 5-foliolate, those on flowering branches usually trifoliolate; the leaflets acuminate, biserrate, sometimes shallowly lobed, varying from nearly glabrous on both surfaces to densely white-tomentose beneath. **Flowers** small, white, in short racemes terminating the flowering branches; sepals lanceolate, hairy, erect or reflexed; petals obovate, white, shorter than sepals, erect or ascending. **Fruit** red, ovoid; the drupelets numerous, finely and thinly tomentose, separating from the receptacle; juicy and sweet. Plate 39, figure 4.

Key to the Varieties of *R. idaeus*

1a Without glands .... *var. idaeus*
1b Glands on stem or at least in inflorescence.
　2a Leaves distinctly grey-tomentose beneath .... **var.** *sacchalinensis*
　2b Leaves green and glabrous or nearly so beneath
　　.. **var.** *peramoenus*

Range: Circumboreal, preferring open clearings and rocky or stony soils. Var. *idaeus* is the European plant that is the cultivated raspberry and that sometimes escapes as a wild plant. Var. *sacchalinensis* (Levl.) Focke (var. *strigosus* [Michaux] Maximowicz) and var. *peramoenus* (Greene) Fernald are native North American varieties. In British Columbia they are rare on the coast but widespread and common across the interior.

Rubus phoenicolasius **Maximowicz**　　　　**Japanese Wineberry**

Tall erect shrub up to 3 m. Stems reddish and densely covered with long, spreading, red, gland-tipped hairs mixed with short, stiff prickles; sticky to handle. **Leaves** trifoliolate; the leaflets broadly ovate, acute to acuminate-tipped, with doubly but bluntly serrate margins, glabrous or nearly so above, densely white-felted beneath; petioles densely glandular-bristly; veins sparsely glandular-bristly beneath. **Flowers** in terminal, dense panicles; sepals densely glandular-hairy, lanceolate, terminally appendaged, enlarging in fruit; petals pink, shorter than sepals, appressed to the stamens. **Fruit** an orange-red, juicy raspberry, separating from the receptacle, edible. Plate 39, figure 5.

Range: Native of Japan; cultivated and rarely escaping on the B.C. coast. The few records from Sidney and Texada Island date from before 1940.

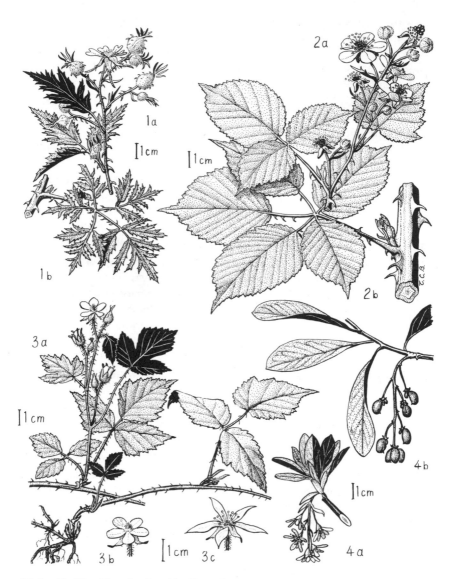

Plate 40 Blackberries and Indian-plum. Figures: (1) *Rubus laciniatus:*
1a, flowering branch; 1b, vegetative stem and leaf; (2) *R. discolor:*
2a, flowering branch; 2b, vegetative stem and leaf; (3) *R. ursinus:*
3a, flowering branch on second-year stem, and rooted tip and first-year
stem and leaf; 3b, pistillate flower; 3c, staminate flower; (4) *Oemleria
cerasiformis:* 4a, inflorescence and emerging leaves, in March; 4b, fruiting
branch, in July.

Rubus ursinus   **Trailing Blackberry**
Chamisso & Schlechtendal var. *macropetalus* (Douglas) Brown

Stems barely woody, weak, long-trailing or arching, rooting at the tips,
somewhat glaucous, with straight to recurved, slender prickles. **Leaves** usually trifoliolate (rarely pinnately 5-lobed) on trailing vegetative stems; the
uppermost leaves on flowering branches simple, sometimes shallowly lobed
and coarsely biserrate; sparsely pubescent, evergreen. Petioles and midveins
beneath with fine prickles; stipules linear, 8-11 mm long, attached to base of
petiole. **Flowers** dioecious, in hairy or prickly small clusters on erect flowering branches from second-year stems; pistillate flowers with petals elliptic
to obovate, 8-11 mm long and very short, non-functional stamens; staminate
flowers with lanceolate petals 12-17 mm long and minute, non-functional
ovaries. **Fruit** a small ovoid blackberry, adherent to the receptacle, black,
juicy and sweet. Plate 40, figure 3. Photo 44.

Range: Southern British Columbia to California in and west of the Coast
and Cascade ranges, and in Idaho. In British Columbia, common on the
coast; becoming abundant in clearings, especially in logged or burnt forest.

Rubus discolor Weihe & Nees   **Himalayan Blackberry**
(*R. procerus* Mueller)

Coarse shrub with high-arching, fluted stems with stout, hooked prickles
along the angles, downy at least when young. **Leaves** palmately 5-folio-
late, except sometimes 1- to 3-foliolate on flowering branches; with ovate
to nearly orbicular leaflets acuminate-tipped and irregularly biserrate on
the margins, greyish-tomentose beneath; petioles and midveins beneath
with short, hooked prickles, variously deciduous to evergreen. **Flowers**
white to pale pink, in terminal tomentose panicles or simple or compound
racemes; sepals densely tomentose; petals ovate, 10-15 mm long; stamens
many, longer than the styles. **Fruit** black and sweet. Plate 40, figure 2.

Range: A cultivated Eurasian species; commonly escaped, especially in
ditches and other wet places in the southwestern part of British Columbia;
now a troublesome weed in some localities.

Note: The nomenclature of the blackberries is in a chaotic state, thanks to
the descriptions of numerous "microspecies" that differ from each other
by almost indistinguishable details. Opinions vary about which is the correct name for this species.

Rubus laciniatus **Willdenow**   **Evergreen Blackberry**
Ragged Blackberry
Cut-leafed Blackberry

Stout, arching shrub, 1-1.5 m or more tall; the dark reddish brown stems fluted or angled, with strong hooked prickles. **Leaves** evergreen, palmately divided into 5 leaflets that may again be pinnately cleft and deeply incised so that they look torn and ragged; sometimes trifoliolate or simple on flowering branches; the leaflets acuminate-tipped and pubescent beneath. Petioles and undersides of the midveins with short hooked prickles. **Flowers** in leafy-bracted terminal corymbs and a few in upper leaf axils; sepals downy and prickly on the back and terminated by narrow leaflet-like appendages; petals spatulate, sometimes apically lobed, 10-15 mm long, white; stamens many, with small anthers, about equalling the many styles. **Fruit** large, globose to ovoid, black when ripe, juicy and sweet. Plate 40, figure 1.

Range: Native of the Old World. A common escape from cultivation on the southern coast of British Columbia, but not aggressive.

Oemleria

Large shrubs with clumped, arching stems; simple, alternate, entire, deciduous leaves; and small, early-deciduous stipules. Flowers in racemes, dioecious; stamens 15; carpels 5, superior. Fruit a group of 1-5 ellipsoid drupes with thin flesh and smooth stones.

Oemleria cerasiformis   **Indian-plum, Oso-berry**
(Torrey & Gray *ex* **Hooker & Arnott) Landon**
(*Osmaronia cerasiformis* **[Torrey & Gray** *ex* **Hooker & Arnott]**
Greene; *Nuttallia cerasiformis* **Torrey & Gray)**

Large shrub up to 4 m tall or, rarely, a small tree. Branchlets brittle-based, rank-smelling when broken, with chambered pith. Buds soft, pale greenish, sometimes pinkish-tinged. **Leaves** narrowly elliptic to oblanceolate, 5-10 cm long, acute at base, blunt to mucronate at apex, entire, thin, sparingly pubescent beneath; the petioles 5-10 mm long. **Flowers** 8-10 mm wide, in 5- to 10-flowered, hanging, glabrous racemes; commonly axillary to basal bud scales of the same buds as give rise to leafy branches. Petals white. Flowering in very early spring, before or with leaf expansion. **Fruit** a plum, 9-12 mm long, bluish black, bloomy and bitter when ripe; 1-5 fruits maturing from each pistillate flower. Plate 40, figure 4. Photo 51.

Range: Southwestern British Columbia to California, between the coast and the Cascade Mountains, on shaded slopes and in clearings at lower elevations. In British Columbia, confined to eastern Vancouver Island from Nanaimo southward and from Howe Sound and Chilliwack southward on the mainland.

Prunus   Cherries and Plums

Trees and shrubs, commonly with smooth, dark bark having horizontally elongated lenticels, as in birches. Twigs with basically alternate buds, but buds commonly clustered at ends of twigs, the buds with many overlapping scales. Leaves alternate, simple, serrate, with petioles that commonly bear a few glands at the junctions with the blades. Flowers in racemes, or sometimes few in umbel-like clusters; sepals 5; petals 5, usually white in wild species; stamens numerous; pistils 1, with elongate style. Fruit a drupe, usually 1-seeded.

Besides our three native species, an increasing number of introduced exotic species have shown an inclination to spread, sometimes only by suckering, into disturbed wild areas on a very local scale. A few of these are mentioned here.

Key to the Species of *Prunus*

1a Leaves evergreen, leathery, glossy, bright green . . . . . *P. laurocerasus*
1b Leaves deciduous, thin, soft, dull-surfaced.
   2a Spiny shrub with widely divergent, pungent spurs; terminal buds absent; leaves commonly without petiolar glands; flowers short-pedicelled, in 2- to 4-flowered axillary umbels; fruit a purple-black plum with a whitish bloom . . . . *P. spinosa*
   2b Unarmed trees; terminal buds present; leaves commonly with petiolar glands.
      3a Inflorescence an elongate raceme of 12 or more flowers
      . . . . *P. virginiana*

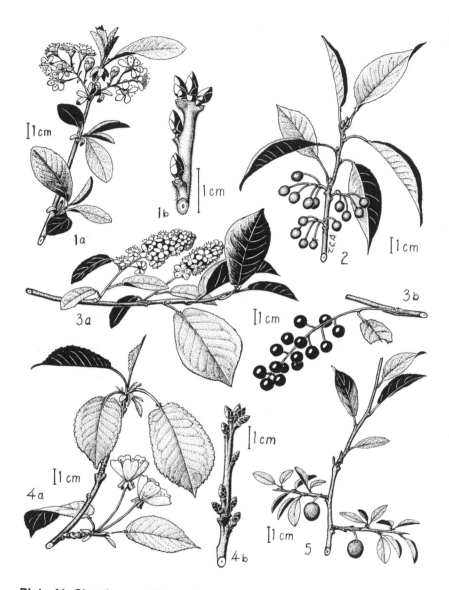

Plate 41 Cherries and Plums. Figures: (1) *Prunus emarginata:* 1a, flow-ering branch; 1b, winter twig; (2) *P. pensylvanica;* (3) *P. virginiana:* 3a, flow-ering branch; 3b, fruiting raceme; (4) *P. avium:* 4a, flowering branch; 4b, winter twig; (5) *P. spinosa.*

3b Inflorescence an umbel or short-peduncled corymb.
4a Leaves elliptic, acute to rounded at tip, 2.5-5 cm long; fruit
an ellipsoid cherry up to 12 mm long, red or purplish
. . . . *P. emarginata*
4b Leaves lanceolate to elliptic, acuminate at tip, longer.
5a Leaves lanceolate, widest below middle, finely and
sharply toothed; fruit globose, up to 12 mm long on a pedicel 1-2 cm long . . . . *P. pensylvanica*
5b Leaves obovate with attenuate tip, widest about the
middle, with coarse, rounded teeth; fruit globose, 20 mm or
more in diameter, on a pedicel 3 cm or more long
. . . .*P. avium*

Prunus pensylvanica L.f.   Pin Cherry, Wild Red Cherry

Small tree up to 12 m tall, or shrub. Branches slender, glabrous, reddish
and shiny. Bark smooth, reddish, with horizontally elongated lenticels.
Buds rounded, tending to be clustered at twig ends. **Leaves** 7-10 cm long,
ovate to oblong-lanceolate, long-pointed, finely and sharply serrate,
glabrous; petioles 1-2 cm long. **Flowers** white, in several 2- to 5-flowered
corymbs on slender pedicels less than 2 cm long; calyx reflexed, tipped
with red; sepals shorter than the hypanthium. **Fruit** globose, 12 mm
across, light red, sour. Plate 41, figure 2.

Associated throughout the range is the variety *P. pensylvanica* var. *saximontana* Rehd., which is usually shrubby; with leaves elliptic-ovate and
less acuminate at apex and corymbs with fewer flowers.

Range: Central British Columbia to the Atlantic coast and southward to
Colorado. In British Columbia, northward at least to the Peace River and
westward to the Burns Lake area.

Prunus emarginata   Bitter Cherry
(Douglas *ex* Hooker) Walpers

Shrub or tree up to 12 m tall, sometimes forming thickets with root suckers. Twigs slightly pubescent at first, becoming glabrous. Bark thin,
smooth, dark reddish, marked with horizontally elongated lenticels. Buds
acute, tending to be clustered at ends of the twigs. **Leaves** 2-6 cm long,
obovate to oblanceolate or elliptic, tapered to the often glandular base;

obtuse to rounded or acute at apex; finely crenate-margined; petioles less than 1 cm long. **Flowers** 5-10 in umbel-like, abbreviated racemes; pedicels less than 1 cm long at flowering time; sepals rounded, entire, shorter than the hypanthium; petals oval, white, 5-7 mm long; stamens about 20. **Fruit** ovoid, 7-10 mm long, red, bitter. Plate 41, figure 1. Photo 45.

Key to the Varieties of *P. emarginata*

1a Leaves and calyx thinly pubescent to glabrous, often shrubs
. . . . **var.** *emarginata*
1b Leaves and calyx densely pubescent; trees . . . . **var.** *mollis*

Range: British Columbia to Montana and California. In British Columbia, abundant, as var. *mollis* (Douglas *ex* Hooker) Brewer, along the coast, including Vancouver Island and the Queen Charlotte Islands, northward to Prince Rupert; and, as var. *emarginata,* from the Cascade and Coast ranges eastward across the interior, especially in the more humid areas, northward at least to Barriere; common in the Kootenay Region.

Where the ranges of this species and *P. pensylvanica* overlap in the interior, intermediate forms, apparently of hybrid origin, are sometimes found.

Prunus virginiana **L.**   Chokecherry

A somewhat smaller tree than *P. emarginata,* often shrubby, producing suckers and forming thickets. Bark finely scaly; young twigs usually densely pubescent, becoming smooth and grey-brown. **Leaves** broadly elliptic or obovate, 5-10 cm long; acute or abruptly acuminate, rounded or slightly cordate at the base; closely serrulate; petioles slender, usually glandular near the top. **Inflorescence** an elongate raceme, 8-13 cm long, sometimes pubescent; terminating short lateral branchlets that often bear tardily deciduous leaves. Sepals short and obtuse; petals short-stalked, round, 4-6 mm long. **Fruit** a globose drupe, usually dark purplish to black, edible but astringent. Plate 41, figure 3. Photo 46.

Key to the Varieties of *P. virginiana*

1a Leaves thin, velvety pubescent beneath. West of Coast and Cascade ranges . . . . **var.** *demissa* **(Nuttall) Torrey**
1b Leaves thicker, glabrous beneath or with few hairs on veins. East of Coast and Cascade ranges . . . . . **var.** *melanocarpa* **(A. Nelson) Sargent**

Range: Transcontinental. Throughout the interior of British Columbia, occasional at the coast.

Prunus avium **L.**   **Gean, Mazzard, Bird Cherry**

Tree up to 25 m tall, of open branching habit; bark dark reddish, smooth with horizontal lenticels. Buds clustered at ends of twigs. **Leaves** 6-15 cm long, obovate, acuminate, serrate, glabrous above, pubescent beneath; petioles 2-5 cm long with two glands at junction with blade. **Flowers** in small sessile umbels; petals 8-15 mm long, white, obovate, with truncate or notched apex. **Fruit** a bright or dark red glabrous drupe. Plate 41, figure 4.

Range: Native to Europe; cultivated and escaped in British Columbia, mainly on the Lower Mainland and southern Vancouver Island.

Prunus spinosa **L.**   **Sloe, Blackthorn**

Tall, profusely branched, spiny shrub, up to 4 m tall, with pubescent young twigs and widely diverging, spine-tipped spurs. Terminal buds absent. **Leaves** narrowly obovate, tapering to a commonly glandless petiole; serrulate, 2-4 cm long. **Flowers** appearing before the leaves, 1 or 2 together, from a bud on a spur, 1-2.5 cm across; white, on a short, glabrous pedicel. **Fruit** a purple to blue-black, pruinose, almost globular plum, 1-1.5 cm in diameter; astringent. Plate 41, figure 5.

Range: Introduced from Europe; occasionally established wild on southern Vancouver Island.

Prunus laurocerasus **L.**   **Cherry-laurel**

Tree up to 6 m tall, or a coarse shrub. Twigs green, glabrous, the buds not clustered at the ends. **Leaves** evergreen, oblanceolate to oblong or elliptic, 8-18 cm long; tapering to base, shortly acuminate at apex; sparingly shallowly serrate; thick, leathery, glossy bright green. **Flowers** white, with circular petals 3-4 mm across, in ascending racemes in the axils of old leaves. **Fruit** a purple-black drupe, 8-15 mm long. Plate 54, figure 1.

Range: Native of southeastern Europe; commonly planted here and used in hedges. Occasionally escaping: records from Vancouver, Port Renfrew and the Gulf Islands.

Malus   Apples

Trees or shrubs with alternate, simple, toothed or lobed leaves. Flowers in corymbs or umbels on lateral spurs; sepals 5; petals 5, rounded, clawed; stamens 15 or more; ovary inferior, the styles united at the base. Fruit a pome, rounded to hollowed at base in ours; the flesh not gritty in texture.

Key to the Species of *Malus*
1a Leaves often shallowly and asymmetrically lobed; spurs diverging perpendicularly from branch, stiff and pungent; flowers 1-2 cm across, white; styles 3-4; fruit purple to brown, 1-2 cm long, shorter than pedicel, rounded at base; the calyx deciduous as fruit ripens
. . . . *M. diversifolia*
1b Leaves never lobed; spurs oblique, not stiff or pungent; flowers 2.5-3 cm across, often pinkish, with 5 styles; fruit globose, 2.5-8 cm long, longer than pedicel, tipped by persisting calyx . . . . *M. domestica*

Malus diversifolia (Bongard) Roemer   Pacific Crabapple
(*M. fusca* [Rafinesque] Schneider)

Tree up to 10 m tall, or shrubby; the trunk with rough, shredding brown bark. Branches with perpendicularly diverging, slender but stiff, sharp but not spiny spur shoots, usually with terminal buds; the young branchlets pubescent. **Leaves** pubescent, becoming glabrous above; ovate to broadly lanceolate, acute to acuminate at apex, 2.5-9 cm long; sharply serrate, often with small lobes on one or both sides. **Flowers** 6-12 in terminal corymbs, up to 25 mm across; pedicels and calyx pubescent; styles 3 or 4. **Fruit** ellipsoid, 10-15 mm long, greenish to purplish or brown. Plate 42, figure 1. Photo 47

Range: Alaska to California along the coast and inland to the Coast and Cascade ranges, in moist, open woods, swampy alluvial flats and lake-shores.

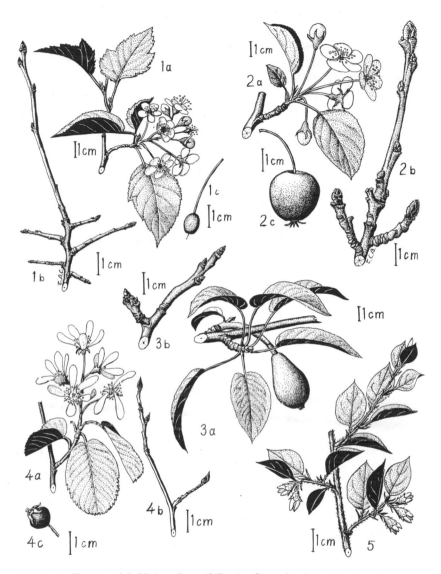

Plate 42 Figures: (1) *Malus diversifolia:* 1a, flowering branch; 1b, winter twig; 1c, fruit; (2) *M. domestica:* 2a, flowering spur-shoot; 2b, winter twig; 2c, fruit; (3) *Pyrus communis:* 3a, fruiting branch; 3b, winter twig; (4) *Amelanchier alnifolia:* 4a, flowering branch; 4b, winter twig; 4c, fruit; (5) *Cotoneaster dielsiana.*

Malus domestica Borkhausen   **Apple**
(*M. sylvestris* [L.] Miller subsp. *mitis* [Wallroth] Mansfeld)

Small tree with brown, eventually flaking bark and usually a coarser, less intricate branching pattern than *M. diversifolia;* the spurs obliquely diverging, thicker and blunter; twigs and foliage more persistently pubescent than in the latter species. **Leaves** never lobed. **Flowers** more abundant than in *M. diversifolia,* white to pinkish, 2.5-3 cm across, with 5 styles. **Fruit** usually globose, at least 2.5 cm long, even on wild trees; hollowed at base and tipped by the persistent calyx. Plate 42, figure 2.

Range: The common cultivated apple, of European origin, this species frequently appears as a wild seedling along fences in farming districts and other settlements. It is frequently found in the Okanagan Valley and on southern Vancouver Island, with other records from Vancouver, Hope and Clearwater (Gerald B. Straley, personal communication).

Pyrus   **Pears**

Deciduous trees and shrubs, generally similar to *Malus,* with simple leaves. Stamens 20-30. Styles 2-5, free. Ovary as in *Malus.* Fruit brownish or greenish, not indented at base, the flesh with grit-cells.

Pyrus communis L.   **Pear**

Trees up to 15 m tall, with round to cylindrical crowns and ascending branches, or coarse shrubs; the blackish bark breaking into small squarish plates. Leaves with long petioles and elliptic blades; woolly when young, but becoming glabrous at least above; the margin serrulate. Flowers in umbel-like corymbs. Petals obovate, white, 10-15 mm long; anthers purple. Fruit 2-12 cm long, tapered to base; the flesh gritty. Plate 42, figure 3.

Range: Native of Europe and western Asia; cultivated here and occasionally escaping on southern Vancouver Island.

Amelanchier Service Berry, Saskatoon Berry

Shrubs or small trees with alternate, deciduous, simple leaves. Flowers white, in terminal racemes. Sepals and petals 5 each; stamens 10-20, borne on the rim of the hypanthium; styles 4-5, united at base in ours. Fruit a berrylike pome crowned with the calyx, with as many locules as styles; the locules incompletely partitioned into 1-seeded half-locules; the fruit bluish black and edible when ripe.

Amelanchier alnifolia Nuttall Saskatoon Berry

Diverse in form: a slender, clumped or colonial shrub; rarely a small tree up to 12 m tall with ascending branches and smooth, grey bark. Twigs reddish brown, smooth; buds reddish, thinly puberulent. **Leaves** broadly elliptic to nearly circular, rounded at base; rounded, truncate or acute at apex; sharply serrate, at least toward the apex; glabrous or thinly hairy and becoming glabrous. **Flowers** variable in size and shape of petals, which generally are narrowly elliptic to oblanceolate. **Fruit** dark bluish to purple, 5-10 mm in diameter; edible when ripe. Plate 42, figure 4. Photo 48.

Range: Alaska to Lake Superior and southward to California, New Mexico and Nebraska. Found in open clearings, often on stony or rocky soils. Widespread all over British Columbia.

Note: Several of the variations shown by this species have been described as species, but the character distinctions tend not to correlate with each other, and intermediate forms are often found. Modern treatments tend to regard these as varieties only. The following key has been adapted from Hitchcock et al. (27, part 3).

Key to the Varieties of *A. alnifolia*
1a Leaves stiff, rather leathery in texture, glabrous or almost so. Petals 15 mm long or less. Top of ovary glabrous. Southern Rocky Mountains and Kootenay Region in British Columbia, to eastern Washington, Montana, Colorado and northeastern California
. . . . **var.** *pumila* **(Nuttall) A. Nelson**
1b Leaves thinner, softer, usually hairy. Top of ovary hairy to glabrous.
2a Petals 12 mm or less long. Top of ovary pubescent.

3a Leaves toothed only near their tips. Styles 4. Southern coast of
British Columbia and Vancouver Island to southwestern
Washington . . **var.** *humptulipensis* **(G.N. Jones) C.L. Hitchcock**
3b Leaves toothed on the apical half. Styles 5. Alaska and interior
of British Columbia to the Dakotas and eastern Oregon
. . . . **var.** *alnifolia*
2b Petals 12-25 mm long.
4a Top of ovary densely pubescent. Petals 12-16 mm long. Sepals
3 mm or less long. Plant sometimes erect and treelike. Alaska to
California along the coast and across the southern interior of
British Columbia to Montana
. . . . **var.** *semiintegrifolia* **(Hooker) C.L. Hitchcock**
4b Top of ovary glabrous to moderately pubescent. Petals 16-25
mm long. Southern interior of British Columbia to Idaho and eastern Oregon . . . . **var.** *cusickii* **(Fernald) C.L. Hitchcock**

Cotoneaster Cotoneasters

Mostly shrubs with unarmed branches; small, scaly buds; and alternate,
simple leaves with entire margins and short petioles. Flowers few in small
corymbs on short lateral branchlets. Stamens about 20. Carpels 2-5, within the hypanthium; connate except on their inner sides; 2-ovuled. Styles
separate. Fruit mealy-fleshy, containing 2-5 nutlets.

A genus of about 50 species in the Old World, many are grown as garden ornamental subjects (52).

Cotoneaster dielsiana Pritzel

Ascending or arching shrub up to 2 m high, with sparsely branching,
pubescent stems. Branches, leaves and calyces with yellowish hairs.
Leaves deciduous, ovate, acuminate, 1-4 cm long, in yellowish-pubescent
petioles 2-4 mm long; the leaf blade deep green and sparsely appressed-
hairy above and pale green and more prominently hairy beneath, with yellowish hairs. **Flowers** 2-7 in small, pubescent cymose corymbs.
Hypanthium funnel-shaped, hairy, its lobes triangular and mucronate.
Petals erect, overlapping, pink; the open flower 8-9 mm long. **Fruit** a
small scarlet pome, 5-6 mm in diameter, containing 3-5 nutlets. Plate 42,
figure 5.

Range: Native of China; planted as an ornamental garden shrub here. Recorded as an occasional escape on Vancouver Island and adjacent islands. *Cotoneaster bullata, C. lactea, C. horizontalis, C. simonsii* and *C. salicifolia* have been reported growing wild in the Vancouver area (Gerald B. Straley, personal communication).

Sorbus   Mountain-ashes, Rowans

Deciduous trees or shrubs; twigs with leaf scars with 5 bundle scars; buds rather large, more or less conical, dark brown in ours, with overlapping scales. Leaves alternate, stipulate, pinnately compound in ours. Flowers white or cream-coloured, small, numerous in terminal flat panicles; stamens many; styles distinct. Fruit an acid-tasting, small pome, usually less than 1 cm across.

Key to the Species of *Sorbus*

1a Small tree with smooth, greyish bark. Buds thickly grey-woolly. Leaflets 11-15, ovate-lanceolate, serrate except for the rounded bases
. . . . *S. aucuparia*
1b Mainly shrubs in British Columbia, rarely small trees. Buds thinly hairy.
   2a Leaflets 9-13, serrate nearly full length, elliptic to lanceolate. Buds with sparse white hairs 0.7-1.5 mm long. Fruit orange to scarlet
   . . . . *S. scopulina*
   2b Leaflets 7-11, serrate above the midpoint; oblong; obtuse to rounded at tip. Buds with short brownish to white hairs less than 0.5 mm long. Fruit crimson red to purplish . . . . *S. sitchensis*

Sorbus scopulina Greene   Western Mountain-ash

Coarse shrub up to 4 m tall. Twigs coarse, sparsely hairy when young; with white hairs. The leaf scars with 5 bundle scars; buds dark brown, somewhat sticky, with sparse, white hairs 0.67-1.5 mm long. **Leaves** 10-16 cm long, with 9-13 leaflets: 11-13 leaflets and early-deciduous stipules in var. *scopulina;* 11 or fewer leaflets and persistent stipules in var. *cascadensis* (G.N. Jones) C.L. Hitchcock. Leaflets lanceolate to narrowly elliptic, acute or acuminate at apex; finely, simply serrate or biserrate nearly to base; commonly shiny above, glabrous or nearly so beneath. **Inflorescence** a flat

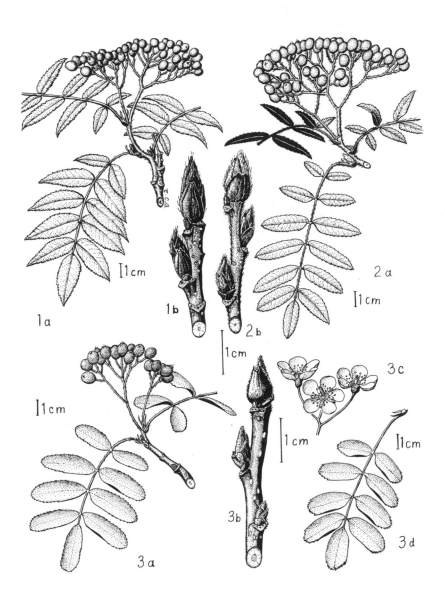

Plate 43 Rowans or Mountain-ashes. Figures: (1) *Sorbus scopulina;*
(2) *S. aucuparia;* (3) *S. sitchensis.* 1a, 2a, 3a, fruiting branches;
1b, 2b, 3b, winter twigs; 3c, 3d, flowers and leaf of *S. sitchensis* var. *grayi.*

corymb 9-15 cm across, with thinly white-pubescent peduncles. **Flowers** 7-9 mm across; hypanthium sparingly hairy; stamens as long as petals (4-5 mm); styles 4 or 5. **Fruit** glabrous, orange to scarlet, shiny. Plate 43, figure 1. Photo 49.

Range: Alaska to Alberta and the Dakotas and southward to New Mexico on low to subalpine wooded slopes. Throughout the mainland of British Columbia; var. *cascadensis* in the Cascade and southern Coast ranges.

Sorbus sitchensis Roemer   Sitka Mountain-ash

Shrub to 3 m tall or, rarely, a small tree. Buds brown, puberulent with short (less than 0.5 mm) brownish to white hairs. **Leaves** 8-18 cm long, with 7-11 leaflets; leaflets oblong, usually dull-surfaced, rounded or obtuse at apex; simply serrate to mid-length or two-thirds of length in var. *sitchensis,* or only close to the apex in var. *grayi* (Wenzig) C.L. Hitchcock (*Sorbus occidentalis* [S. Watson] Greene). **Inflorescence** convex, less than 10 cm across, with brown-hairy peduncles. **Flowers** up to 11 mm across; stamens shorter than petals; styles 3-5. **Fruit** glabrous, crimson red to purplish, often glaucous, not shiny. Plate 43, figure 3. Photo 56.

Range: Alaska to California, from the coast eastward to Montana. In British Columbia, across the interior northward to Hazelton and Mount Robson; var. *grayi* in the Cascade and Coast ranges, at least as far north as Telkwa and on Vancouver Island and the Queen Charlotte Islands, at subalpine and alpine elevations.

Sorbus aucuparia L.   Rowan, European Mountain-ash

Smallish tree up to 15 m tall, with smooth, greyish bark. Twigs coarse, hairy when young. Buds reddish brown, more or less covered with dense, grey pubescence. **Leaves** 10-25 cm long, with 11-15 leaflets; the leaflets lanceolate to elliptic, simply serrate for about ¾ their length, pubescent beneath at least when young. **Inflorescence** 10-15 cm across, the peduncles greyish-hairy when young. **Flowers** 8-12 mm across, the hypanthium pubescent, the petals almost circular; stamens 20, as long as petals; styles 3 or 4. **Fruit** about 10 mm across, red to deep orange, with few remaining hairs. Plate 43, figure 2.

Range: Native to Europe; commonly cultivated here and escaping into woods on Vancouver Island and the southern coast and in the interior as far north as Revelstoke. Often found established on old stumps.

Crataegus   Hawthorns

Small trees and shrubs with rough, scaly or longitudinally grooved bark. Branchlets slightly zigzag, armed with axillary thorns and with buds beside the thorns. Leaves alternate, simple, deciduous, toothed to deeply lobed; stipules linear to lanceolate, often deciduous early. Flowers white or, occasionally, pink; in broad, terminal corymbs; calyx cup bell-shaped, united with the ovary, bearing 5 sepals, 5 petals and numerous stamens on its rim; styles 1-5. Fruit a small pome surmounted by the remains of the sepals and containing 1-5 grooved stones, each containing 1 seed.

Key to the Species of *Crataegus*
1a Spines 40-70 mm long; fruit red .... *C. columbiana*
1b Spines 8-25 mm long.
    2a Leaves deeply pinnately lobed, 15-20 mm long; fruit red; style and
    stone 1 .... *C. monogyna*
    2b Leaves serrate or biserrate, sometimes shallowly lobed toward
    apex, 25-75 mm long. Fruit black. Styles and stones 3-5.
        3a Flowers 14-16 mm wide, with 10 stamens. Branches reddish
        brown. Leaves broadly elliptic to obovate; commonly some leaves
        shallowly lobed toward apex. Thorns 13-25 mm long
        .... *C. douglasii*
        3b Flowers 12-15 mm wide, with 15-20 stamens. Branches pale
        greyish brown. Leaves elliptic or narrowly obovate, sometimes
        apically lobed. Thorns 8-12 mm long .... *C. suksdorfii*

Crataegus douglasii **Lindley**   **Douglas Hawthorn**
                                        Black Hawthorn

Big, coarse shrub or small tree up to 12 m tall, with rough, scaly, brown bark. Branchlets dark reddish brown, with thorns 12-25 mm long, rarely unarmed. **Leaves** 25-75 mm long, broadly obovate, biserrate above the cuneate base, those of young shoots often shallowly lobed; glabrous or pubescent on the midvein; petioles sometimes glandular. **Flowers** usually many in flat corymbs, about 15 mm wide, sepals shortly triangular, more or less villous toward apex; petals orbicular, white; stamens 10, with pink anthers; styles 4 or 5. **Fruit** black, glabrous. Plate 44, figure 1.

Range: Widespread in British Columbia northward to about lat. 55°N, commonly along streams. Mainly in the interior, but also on the southern coast, including southern Vancouver Island. Eastward to Alberta and southward to Oregon and Idaho, with an isolated area in Ontario and Michigan.

Crataegus suksdorfii (Sargent) Kruschke   **Black Hawthorn**
(*C. douglasii* Lindley   **Suksdorf's Hawthorn**
var. *suksdorfii* Sargent)

Small tree or big shrub, closely similar to *C. douglasii;* with branchlets usually pale greyish brown; the thorns stout, 8-12 mm long. **Leaves** similar to those of *C. douglasii,* but tending to be narrower on the average. **Flowers** 12-15 mm wide, with usually 20 but sometimes only 15 stamens; styles 5; flowering in June. **Fruit** black, glabrous. Plate 44, figure 2. Photo 52.

Range: *C. suksdorfii* is found mostly in moist, low-lying sites and along stream banks. It is distributed mainly west of the Cascade and Coast ranges, northward to the Skeena River and including the Queen Charlotte Islands and Vancouver Island, with outlying occurrences in the moister parts of the southern interior and along the Fraser River in the Cariboo Region. It is also found in Idaho, Washington, Oregon and coastal California.

Note: Stamen number is the preferred differentiating characteristic between *C. douglasii* and *C. suksdorfii.* However, in the absence of flowers, the colour of the branches and branchlets appears to be the most reliable difference between them. The range of variation in leaf form is so very wide in each species that it is an unreliable character to use to distinguish between them.

Brunsfeld and Johnson (8) have presented a good argument for regarding this as a species distinct from *C. douglasii.* A significant part of their argument is based on the apparent mutual reproductive isolation of these species resulting from differences in their chromosome complements ($2n$ = 68 for *C. douglasii,* 34 for *C. suksdorfii*).

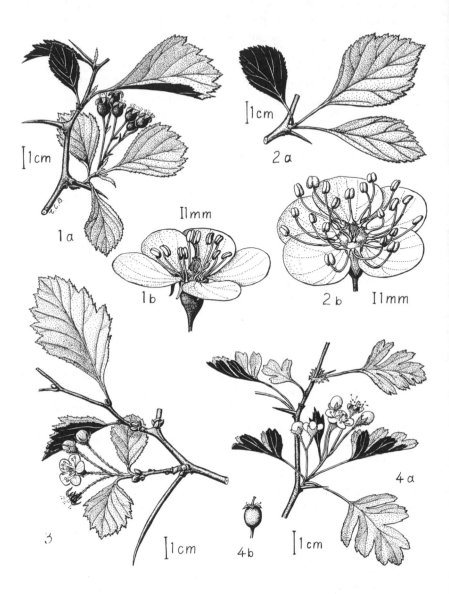

Plate 44 Hawthorns. Figures: (1) *Crataegus douglasii;* (2) *C. suksdorfii;* (3) *C. columbiana;* (4) *C. monogyna.* 1b, 2b, flowers; 4b, fruit.

It has been observed that where *C. douglasii* and *C. suksdorfii* grow in mixture, as at Oak Bay, near Victoria, *C. douglasii* flowers earlier and has finished flowering at the end of May, about the time the adjacent *C. suksdorfii* individuals are starting – another mechanism of reproductive isolation.

Crataegus columbiana Howell   Columbia Hawthorn

Coarse shrub with arching branches, or, rarely, a small tree up to 5 m tall. Branchlets medium brown, with strong thorns 4-7 cm long. **Leaves** short-petioled, obovate, biserrate to shallowly lobed with rather divergent teeth; stipules small, promptly deciduous. **Inflorescence** a flat or convex corymb, with distinctly tomentose peduncles and pedicels. **Flowers** 1.5 cm wide; ovary and hypanthium tomentose outside and inside; sepals lanceolate; petals white, broadly obovate; stamens 10, or sometimes fewer; styles 3 or 4. **Fruit** red. Plate 44, figure 3.

Range: British Columbia to California. In British Columbia, scattered but uncommon in the eastern interior, northward to the Peace River and westward to the Okanagan Valley and to the Fraser River at Prince George; generally in open areas near streams.

Crataegus monogyna Jacquin   English Hawthorn
        May Tree

Small tree up to 10 m tall, with fluted trunk and dark grey-brown, scaly bark. Branchlets greyish to reddish brown, with short, stout thorns mostly around 1 cm long. **Leaves** deeply pinnately lobed, divided more than halfway to the midvein, with rounded, sparingly toothed lobes. **Flowers** in flattish clusters; the ovary and hypanthium pubescent. Petals orbicular, white; sometimes pink in cultivated varieties but rarely so in wild trees. Stamens 20; styles 1. **Fruit** red, with a single stone. Plate 44, figure 4.

Range: Native of Europe; commonly planted; escaping and becoming common and weedy on waste ground, preferentially where moist and open, on southern Vancouver Island and Saltspring Island, and at Vancouver and Crawford Bay (Kootenay Lake) on the mainland.

Hybrids between *C. monogyna* and *C. suksdorfii* have been reported for Oregon (40). The hybrids have rather rounded leaf lobes of intermediate depth and black fruit. Vegetatively, they resemble the European *C. laevigata* (Poiret) DC. (*C. oxyacanthoides* Thuill.), which, however, has red fruit.

This hybrid may be looked for in British Columbia where the parent species grow. *C. laevigata* is occasionally planted here for ornament, but there is yet no record of it naturalizing itself.

Leguminosae (Fabaceae)

Pea Family

Herbs, shrubs, vines or trees; the woody members frequently spiny. Leaves alternate, usually compound, with stipules. Flowers with 5 petals, bilaterally symmetrical in all of ours. The upper petal, called the *standard,* enclosing the others in the bud; the 2 lowest united to form the *keel,* which encloses the 10 united stamens and 1 pistil; the 2 laterals, called the *wings,* variously covering the keel or spreading. Fruit a pod (*legume*), usually splitting lengthwise into 2 spirally diverging valves. All the woody species of this family in British Columbia are of exotic origin. The family is treated in full for this province by Taylor (59).

Key to Woody Genera of Leguminosae
1a Plant spiny or at least with some thorns.
2a Densely branched shrub with green twigs that are spines. Leaves simple, spinelike, inconspicuous. Flowers yellow . . . . *Ulex*
2b Prolifically suckering tree with paired brown, stout thorns at nodes of brownish branches. Leaves mostly 18 cm long or more, pinnately compound with elliptic leaflets. Flowers white, rarely pink
. . . . *Robinia*
1b Plant not at all spiny. Flowers yellow.
3a Branchlets ascending, obliquely or nearly parallel; bright green, glabrous. Leaves small, inconspicuous, of 1-3 leaflets, often shed early.
4a Twigs angular in section, slender, nearly parallel . . . . . . *Cytisus*
4b Twigs cylindrical in section, stouter, obliquely ascending
. . . . *Spartium*
3b Branchlets spreading and ascending, grey-hairy at least when young. Leaves ample, palmately compound or trifoliolate, persisting at least through summer. Flowers in racemes.
5a Small tree with trifoliolate leaves. Racemes hanging in ours
. . . . *Laburnum*
5b Shrub with palmately compound leaves with many leaflets. Flowers in erect racemes . . . . *Lupinus*

Robinia

Deciduous trees and shrubs. Buds small; the axillary buds hidden in pits under the leaf bases until they shoot in spring; the occasional supra-axillary buds minute, on the twig surface above. Terminal buds absent. Stipules commonly but not always replaced by paired thorns. **Leaves** alternate, pinnately compound; the leaflets entire. **Flowers** in hanging racemes, pealike in form. **Fruit** a flat, several-seeded pod, splitting lengthwise into 2 valves.

Robinia pseudo-acacia L.
Black Locust
False Acacia

Tree to 25 m tall, with often crooked, spreading branches. Bark yellowish brown, becoming thick with age, with longitudinal, interlacing ridges producing a ropy appearance. Twigs and foliage lightly pubescent when young, becoming glabrous later; with pairs of stout thorns in place of stipules, especially on vigorous shoots. **Leaves** with 7-19 leaflets that are elliptic, shortly stalked, rounded or mucronate at tips, entire. **Flowers** white, fragrant, about 2 cm long, in a raceme 10-20 cm long. **Fruit** a linear to oblong, flat, 3- to 10-seeded pod. Plate 45, figure 1.

Range: Native of the east-central United States; widely cultivated and escaped, commonly spreading by root suckers, on southern Vancouver Island and across the southern interior of British Columbia.

Laburnum
Laburnum

Small, unarmed trees or shrubs with smooth bark. Leaves trifoliolate. Flowers in racemes, yellow. Calyx shallowly 2-lipped and toothed. Stamens joined into a tube. Anthers of 2 sizes. Fruit a thick pod, somewhat dilated around the poisonous seed.

Laburnum anagyroides Medicus

Golden-chain Laburnum

Tree up to 7 m tall, with smooth, yellowish bark. Twigs appressed-pubescent. **Leaves** with 3 elliptic, entire leaflets, obtuse to mucronate at tips, 4-8 cm long; the petiole can be longer or shorter. **Flowers** about 2 cm long, in racemes that are 10-20 cm long and pendulous. **Fruit** a rather curved pod, 5 cm or more long, poisonous. Plate 45, figure 2.

Range: Native of Europe, commonly planted here and occasionally escaping on southern Vancouver Island.

Lupinus

Lupins

Mostly herbaceous plants with palmately compound leaves, and flowers in terminal, erect racemes. The pealike flowers with deeply 2-lipped calyx, the wing petals coherent and enclosing the sickle-shaped keel and the 10 stamens all joined together by their filaments. Fruit a hairy pod. One shrubby species in British Columbia.

Lupinus arboreus Sims

Bush Lupin

Shrub 1-2 m tall. Stems and branches spreading and ascending. Branchlets and leaves greyish and hairy. **Flowers** pale yellow, roughly whorled on the axis of the raceme. **Fruit** a hairy pod. Plate 45, figure 3.

Range: Native of California; introduced and established on seaward slopes and bluffs on southern Vancouver Island and the Gulf Islands.

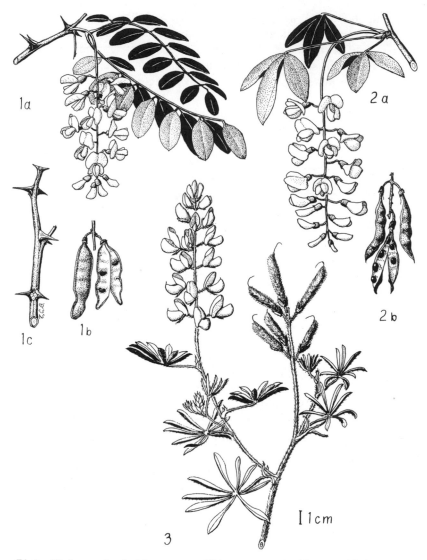

Plate 45 Large-leafed Legumes. All to same scale. Figures: (1) *Robinia pseudo-acacia:* 1a, flowering branch; 1b, fruit, closed and open pods; 1c, winter twig; (2) *Laburnum anagyroides:* 2a, flowering branch; 2b, fruit, closed and open pods; (3) *Lupinus arboreus.*

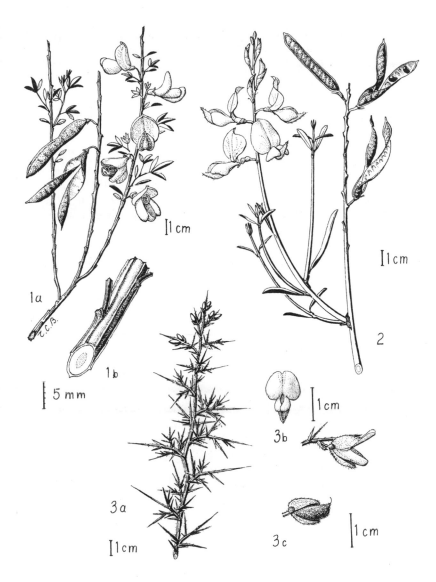

Plate 46 Broom and Gorse. Figures: (1) *Cytisus scoparius:* 1a, branch with flowers and fruit; 1b, section of twig; (2) *Spartium junceum,* branch with flowers and fruit; (3) *Ulex europaeus:* 3a, branch; 3b, flowers; 3c, fruit.

Cytisus   Brooms

Shrubs with ascending, green, angular branches. Leaves small, simple or trifoliolate, often shed early. Flowers axillary, scattered along the twigs; the calyx with 2 short, broad lips; stamens united. Pod linear, splitting lengthwise into 2 valves.

Cytisus scoparius (L.) Link   Scotch Broom

Shrubs with slender, nearly glabrous, longitudinally angled branchlets, ascending nearly parallel to each other. **Leaves** usually less than 1.5 cm long, with 1 or 3 leaflets, often deciduous early. **Flowers** usually yellow, occasionally orange, solitary or in pairs along the branchlets in the axils of old leaves. **Fruit** a dark grey pod, splitting into two valves that spiral apart. Plate 46, figure 1.

Range: Native of Europe. Naturalized and spreading in southern British Columbia. Common on the coast south of lat. 50°N and occasionally in the South Kootenay region, as at Nelson and Creston. It readily invades openings but not closed forest stands, since it does not tolerate shade.

Spartium

Shrubs of broomlike form, with slender, obliquely ascending, almost leafless, green branches. Leaves simple, entire, without stipules. Flowers in erect terminal racemes. Calyx split above: with one 5-toothed lower lip. Wing petals shorter than the acuminate keel, adherent at their bases to the stamen tube. Fruit a multiseeded linear pod.

Spartium junceum L.   Spanish Broom

Shrub up to 3 m or more tall, with mostly bare, green, smoothly cylindrical, rushlike branchlets, bearing minute appressed, white hairs when young. **Leaves** widely scattered, alternate to sometimes nearly opposite, oblanceolate to linear, 1-4 cm long, glabrescent. **Flowers** yellow, 2.5-2.8 cm long, fragrant. **Fruit** a linear pubescent pod, 5-10 cm long. Plate 46, figure 2.

Range: Native to the Mediterranean countries. Introduced in British Columbia; escaping sparingly on southern Vancouver Island in scrub near the sea.

Ulex   Gorse

Shrubs forming dense, spiny thickets. Twigs modified into angular, green, branching, sharp spines. Leaves small, simple, spinelike. Calyx divided into 2 lips (upper and lower). Stamens united into a tube.

Ulex europaeus L.   Gorse, Whin, Furze

Dense, spreading shrub; the branchlets densely branched, longitudinally grooved, green and thinly hairy when young; all branch twigs spine-tipped. **Leaves** simple, linear, usually a centimetre or less long; reduced to a flattish spine without stipules. **Flowers** 1 or 2 in leaf axils, yellow, fragrant. Calyx prominent and deeply 2-lipped, a little shorter than the petals; hairy. Wings longer than the keel and covering it. **Fruit** a short pod about 15 mm long, hairy. Plate 46, figure 3.

Range: Native of Europe; naturalized on Vancouver Island, the Queen Charlotte Islands and the southern coast of the mainland.

Anacardiaceae　　　Cashew Family

Trees or shrubs, evergreen or deciduous, with a milky or resinous juice. Leaves mostly compound and large, alternate, without stipules. Flowers small, in terminal or axillary clusters, greenish; calyx 5- to 7-parted; petals small, 5-7; stamens alternate with the petals and opposite the sepals; ovary superior, 3-styled, sometimes 1-celled and 1-ovuled. Fruit usually a small, semi-woody, globular drupe containing 1 seed.

Rhus　　　Sumac

Shrubs, vines or small trees; ours with pinnately compound or trifoliolate leaves and hairy buds. Stipules absent. Flowers dioecious or perfect; sepals, petals and stamens usually 5; petals and stamens inserted under the edge of a disc lining the hypanthium. Fruit a small drupe.

Key to the Species of *Rhus*

1a　Leaflets 11-31. Fruits red, in terminal panicles. Terminal bud absent.
　　2a　Branches and leaf stalks glabrous .... *R. glabra*
　　2b　Branches and leaf stalks densely hairy .... *R. typhina*
1b　Leaflets 3, rarely 5. Fruits white, in mainly axillary racemes.
　　Terminal bud present.
　　3a　Leaflets acuminate-tipped .... *R. radicans*
　　3b　Leaflets obtuse to rounded at tips .... *R. diversiloba*

Rhus glabra L.　　　Smooth Sumac

Big coarse shrub, spreading by suckers; or sometimes a small tree up to 5 m tall. Bark reddish brown, smooth. Twigs glabrous, greenish to reddish, often with a bloom. Axillary buds almost surrounded and concealed by the leaf bases. Terminal bud absent. **Leaves** 20-35 cm long, pinnately compound with 11-31 leaflets that are lanceolate, acuminate, 5-10 cm long and simply serrate; green above, and paler beneath with a bloom; turning a deep red in autumn. **Flowers** small, greenish, in a finely pubescent terminal panicle 10 cm or more long. **Fruit** dark red, with very fine, short, velvety puberulence. Plate 47, figure 1. Photo 53.

Range: Over most of the United States, and in southern British Columbia, Ontario and Quebec. In British Columbia, in the dry southern interior, northward on the Fraser River as far as Williams Lake, and in the Okanagan and Kootenay regions, on rocky soils or alluvial gravels, often in association with Ponderosa Pine and Douglas-fir.

Rhus typhina L. in L. & Torner   Staghorn Sumac

Similar in habit to *R. glabra*. A small tree up to 8 m tall, or a big, coarse colonial shrub spreading by suckers. Branches, twigs and leaf stalks densely pilose: covered with soft perpendicular brownish hairs; axillary buds concealed by the leaf bases; terminal bud absent. **Leaves** up to 40 cm long, pinnately compound, with 11-31 lanceolate, simply serrate leaflets, green above and paler beneath; turning scarlet in autumn. **Flowers** minute, greenish; in dense, terminal, hairy panicles. **Fruit** densely pilose, the hairs longer than those of *R. glabra*. Plate 47, figure 2.

Range: Native to eastern North America from Quebec to Georgia on open rocky sites. It is commonly planted in British Columbia as an ornamental plant and has been reported as escaping in Saanich and on Saltspring Island.

Rhus diversiloba Torrey & Gray   Poison-oak

Shrubs up to 2 m tall; erect or climbing with aerial rootlets. Stems and branches yellowish brown, **contact-poisonous**; terminal bud present, axillary buds not concealed by the leaf bases. **Leaves** normally trifoliolate; leaflets 20-75 mm long, variable in form; obovate to ovate, rounded or obtuse at apex, the margin sinuately lobed to entire; the lateral leaflets sessile or on stalks 1-3 mm long. **Flowers** minute, green, in small axillary racemes or panicles. **Fruit** a small white globose glabrous drupe. Plate 47, figure 3.

Range: British Columbia to Mexico, mainly between the coast and the Cascade Mountains. In British Columbia, on the southern coast about Howe Sound, on Vancouver Island northward to Nanaimo and in the Gulf Islands northward to Texada Island; preferring rocky, semi-open places.

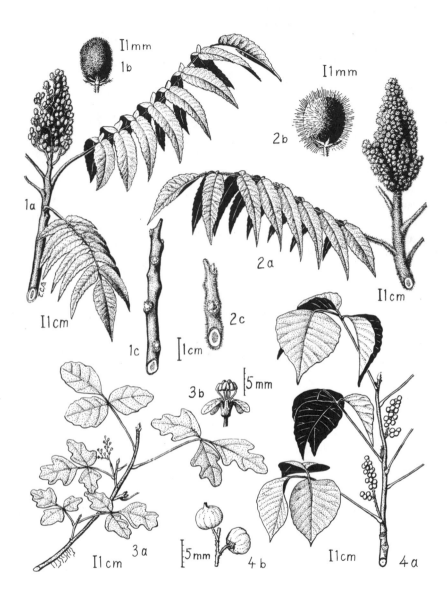

Plate 47 Sumac and Poison-ivy. Figures: (1) *Rhus glabra:* 1a, fruiting branch; 1b, fruit; 1c, winter twig; (2) *R. typhina:* 2a, fruiting branch; 2b, fruit; 2c, winter twig; (3) *R. diversiloba:* 3a, branch; 3b, flower; (4) *R. radicans:* 4a, fruiting branch; 4b, fruits.

Rhus radicans L. var. *rydbergii* (Small) Rehder   Poison-ivy

Shrub up to a metre or more tall, erect or prostrate, with yellowish brown stems, **contact-poisonous**; terminal bud present; axillary buds often not completely concealed by leaf bases. **Leaves** trifoliolate; leaflets 50-80 mm long, broadly ovate; acuminate at apex, broadly rounded at base; coarsely round-toothed or entire; glabrous above, thinly hairy beneath; the lateral leaflets on stalks 3-5 mm long; the leaves turning scarlet in autumn. **Flowers** in simple or branching racemes 5-10 cm long, from the axils of the upper leaves; petals oblong, greenish, longer than the sepals. **Fruit** a small, white, glabrous, ribbed, persistent drupe. Plate 47, figure 4.

Range: From the Fraser River in British Columbia eastward over much of the United States, and in southern Ontario, where var. *radicans,* which often climbs trees, also occurs. In British Columbia, it is found mainly in the dry southern interior: on the Fraser River between Hope and Macalister (which is between Williams Lake and Quesnel), to Adams River in the South Thompson River valley, to Canal Flats in the Kootenay Valley, and around the hot springs at Fairmont and Ainsworth. Var. *radicans,* found in eastern North America, is a high-climbing vine.

Note: *Rhus radicans* and *R. diversiloba* are placed by some authors in the genus *Toxicodendron,* distinct from *Rhus.* They exude an oil that can cause intense irritation of the skin from contact.

Aquifoliaceae   Holly Family

Trees or shrubs with small, 4- to 9-parted flowers generally clustered in leaf axils; staminate and pistillate flowers are often on separate plants. Stamens as many as petals; ovary free from calyx. Fruit a 4- to 9-seeded berry.

Ilex   Holly

Trees and shrubs with alternate simple leaves and small axillary white flowers that are variously staminate, pistillate (often dioecious) or perfect; petals distinct or coherent; stamens adherent to the petal bases. Fruit a berrylike drupe containing 4-9 nutlets.

Ilex aquifolium L.   English Holly

Big shrub to small tree with smooth grey bark and white wood. **Leaves** tough and leathery, evergreen, glossy, with undulate margins bearing a few spiny teeth on each side. **Flowers:** staminate and pistillate commonly on separate plants. **Fruit** red berries in clusters, persistent through the winter. Plate 48, figure 4.

Range: Native of Europe; cultivated in southwestern British Columbia and escaped locally into woods around Vancouver and Victoria.

Celastraceae   Staff-tree Family

Shrubs, small trees or vines. Leaves simple, and opposite and serrulate in our species. Flowers small, radially symmetrical, in cymes with 1 to a few flowers; calyces with 4 or 5 lobes, or parted; stamens alternating with the petals and attached to the edge of a broad disc; ovary superior, with 2 or 3 chambers. Fruit a capsule in ours, or a follicle. Seeds with fleshy appendages.

Key to the Genera of Celastraceae
1a Leaves evergreen, 1-3 cm long; petals 4, red . . . . *Paxistima*
1b Leaves deciduous, 5-10 cm long; petals 5, greenish and purplish-mottled . . . . *Euonymus*

Paxistima (*Pachistima*)

Low shrubs with evergreen, opposite, serrulate, glabrous, leathery leaves. Flowers very small; sepals, petals and stamens 4 each. Fruit a 2-chambered capsule.

Paxistima myrsinites (Pursh) Rafinesque   False-box
(*Pachistima myrsinites* [Pursh] Rafinesque)

Evergreen, low shrub, 30-80 cm tall, densely branched with ascending slender branches. **Leaves** 5-30 mm long, elliptic to lanceolate; tapered to base, acute to obtuse at apex; the margins serrulate; both surfaces glabrous and rather shiny. **Flowers** in small few-flowered axillary clusters; the petals minute, usually brick-red, rarely yellowish. Plate 48, figure 2. Photo 54.

Range: In woods on rocky soils; from central British Columbia to California. In British Columbia, from Vancouver Island to the Rocky Mountains and northward in the interior at least to Smithers.

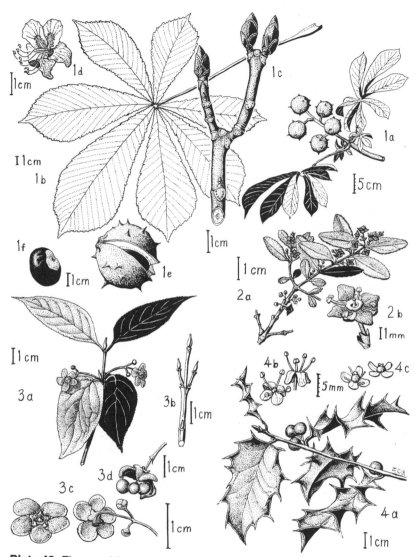

Plate 48 Figures: (1) *Aesculus hippocastanum:* 1a, fruiting branch; 1b, leaf; 1c, winter twig; 1d, staminate flower; 1e, fruit; 1f, seed; (2) *Paxistima myrsinites:* 2a, flowering branch; 2b, flower; (3) *Euonymus occidentalis:* 3a, flowering branch; 3b, winter twig; 3c, flowers; 3d, fruit; (4) *Ilex aquifolium:* 4a, fruiting branch; 4b, staminate flowers; 4c, pistillate flowers.

Euonymus   Spindle-trees

Flowers with 5 petals. Fruit a capsule opening to expose seeds enveloped in coloured, fleshy arils.

Euonymus occidentalis **Nuttall** *ex* **Torrey**   Burning-bush, Western Wahoo

Tall, open shrub, 2-5 m tall. Branches glabrous. Twigs glabrous, green, longitudinally ribbed. Buds acuminate-tipped. **Leaves** ovate to lanceolate, acuminate, finely crenate-serrulate, thin and deciduous. **Flowers** in stalked 3-flowered groups in leaf axils; petals mottled greenish and purplish, with pale margins; anthers longer than the short, stout filaments, opening across the top; ovary sitting in a pit in the centre of the disc. **Fruit** opening by 3 valves to expose seeds covered by orange-red, fleshy arils. Plate 48, figure 3.

Range: Coastal areas of the Pacific states. Very rare in British Columbia; found near Courtenay, in alluvial woods.

Aceraceae   Maple Family

Trees or shrubs with opposite, simple or sometimes compound leaves. Flowers small; variously perfect, perfect and unisexual on the same plant, or dioecious, in clusters; carpels 2, rarely 3, united below. Fruit splitting at maturity into a pair of winged, 1-seeded carpels (samaras).

Acer   Maples

Deciduous trees and shrubs. Buds with overlapping scales or with two outer scales only. Leaves simple and palmately lobed or sometimes pinnately compound, but without stipules. Flowers dimly coloured or greenish; calyx usually 5-lobed; petals as many as calyx lobes, or absent. Stamens 3-12, inserted at the edge of a disc; carpels 2 or 3, with 2 ovules each; styles 2, united at base. Fruit 2 or 3 long-winged, 1-seeded carpels, separating at maturity. Seedling with 2 elongate cotyledons.

Key to the Species of *Acer*
1a Leaves pinnately compound .... *A. negundo*
1b Leaves simple and palmately lobed.
   2a Leaf lobes lobed to almost entire, not serrate.
      3a Leaf blade indented more than halfway to base between lobes. Central lobe 10 cm or more long, flowers in long, hanging panicles. Fruit with stiff hairs .... *A. macrophyllum*
      3b Leaf blade indented less than halfway to base. Central lobe less than 10 cm long. Flowers in short, broad, stiffly projecting or erect panicles. Fruit glabrous .... *A. platanoides*
   2b Leaf lobes serrate-margined.
      4a Leaf blade with 3-5 lobes, rarely divided into 3 leaflets. Petiole usually longer than leaf blade. Fruit wings spreading by about 90° .... *A. glabrum*
      4b Leaf blade with 7-9 lobes. Petiole shorter than leaf blade. Fruit wings spreading by 180° .... *A. circinatum*

Acer macrophyllum **Pursh**   **Broadleaf Maple**

A large tree up to 30 m tall, with widely spreading branches. Bark light brown, closely and shallowly grooved; in shady situations often covered with epiphytic plants. Buds with several pairs of scales; terminal bud large; leaf scars with 5 or more bundle scars. **Leaves** 15-30 cm or more wide, deeply and sinuately 5-lobed; the lobes variously notched or almost entire; dark green above, pale green, and slightly pubescent beneath when young. **Flowers** greenish to straw-coloured; sepals and petals equal or almost so; stamens with hairy filaments; both perfect and staminate flowers borne together in many-flowered, hanging racemes appearing before the leaves are expanded. **Fruit** with wings about 5 cm long, spreading at about 90° and with bristly, yellowish hairs on the nutlets. Plate 49, figure 4. Photo 55.

Range: Southwestern British Columbia to southern California along the coast and inland to the Cascade Range, in open coniferous stands and on alluvial flats, and often a pioneer tree on logged-off lands. In British Columbia, north on Vancouver Island to Kyuquot Inlet and the Tahsis River, on the mainland coast at least to Port Neville, and inland to Seton Portage near Lillooet, Siska in the Fraser Canyon and the Skagit River valley near Hope.

Acer circinatum **Pursh**   **Vine Maple**

Small tree or big spreading shrub, 3-10 m tall. The branches slender with glabrous twigs, light green, sometimes tinged with red. Leaf scars with 3 bundle scars; buds covered by the outer pair of scales, shiny red, fringed with white hairs; terminal bud usually absent. **Leaves** 5-13 cm across; circular, cordate-based; shallowly 7- to 9-lobed, the lobes acute-tipped; serrate; the surface glabrous above, puberulous beneath, becoming glabrous; in autumn turning bright scarlet or yellow; petioles about 4 cm long. **Flowers** in small, loose terminal corymbs, perfect or staminate; sepals red, petals white. **Fruit** glabrous, the wings spreading at 180° apart, up to 4 cm long; the nutlet ribbed. Plate 49, figure 3.

Range: Southwestern British Columbia to California along the coast and inland to the eastern slope of the Cascade Range. In British Columbia, in moist woods of the coast across from Vancouver Island and inland to

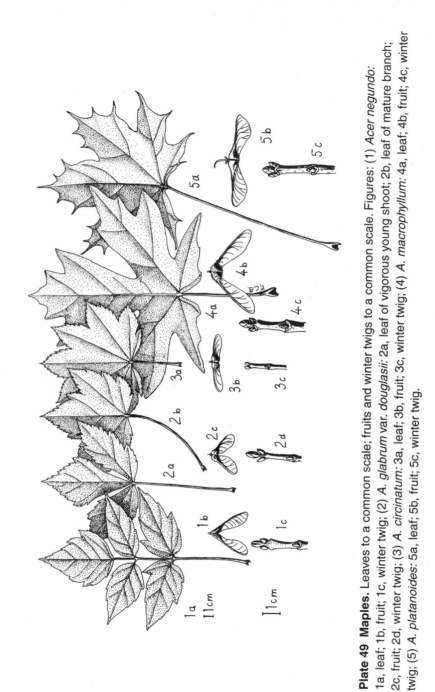

Plate 49 Maples. Leaves to a common scale; fruits and winter twigs to a common scale. Figures: (1) *Acer negundo:* 1a, leaf; 1b, fruit; 1c, winter twig; (2) *A. glabrum* var. *douglasii:* 2a, leaf of vigorous young shoot; 2b, leaf of mature branch; 2c, fruit; 2d, winter twig; (3) *A. circinatum:* 3a, leaf; 3b, fruit; 3c, winter twig; (4) *A. macrophyllum:* 4a, leaf; 4b, fruit; 4c, winter twig; (5) *A. platanoides:* 5a, leaf; 5b, fruit; 5c, winter twig.

Manning Provincial Park; almost absent from Vancouver Island, the occurrences on the San Juan and Robertson rivers suspected of being from introductions.

Acer glabrum Torrey   Douglas Maple
var. *douglasii* (Hooker) Dippel   Rocky Mountain Maple

Small tree up to 10 m tall, or big shrub with slender ascending reddish smooth branches; twigs dark red; leaf scars with 3 bundle scars. Buds with 1 or 2 pairs of visible scales; the terminal bud present, slightly raised above the adjacent lateral buds. **Leaves** 6-10 cm wide, 3- to 5-lobed or deeply cleft, occasionally divided into 3 leaflets; coarsely biserrate. **Flowers** in corymbs appearing with the leaves; unisexual, the staminate and pistillate flowers usually on separate trees; sepals and petals almost alike, yellowish green; staminate flowers with rudimentary pistil, pistillate flowers with small, non-functional stamens. **Fruit** glabrous, with wide wings about 25 mm long, diverging at about 90°. Plate 49, figure 2.

Range: Alaska to Alberta, Montana and California (var. *glabrum* from Montana to New Mexico and Arizona) in moist hollows and on rocky slopes. In British Columbia, from Vancouver Island to the Rocky Mountains and over most of the interior of the province northward to the Peace River and adjacent to the Alaskan border to Rainy Hollow (lat. 59°32'N, long. 136°30'W).

Acer negundo L.   Manitoba Maple, Box-elder

Long-limbed tree with grey, smooth to shallowly furrowed bark. Twigs pale green with a white bloom; buds with 1 or 2 pairs of hairy scales. **Leaves** pinnately compound, of 5-7 sparingly toothed to shallowly lobed leaflets; light green above, greyish green beneath. **Flowers** of 2 kinds, the staminate and pistillate on separate trees; minute, without petals; staminate flowers on long stalks in loose clusters, pistillate flowers arranged along a central stalk; flowering before or with leaf development in early spring. **Fruit** a pair of samaras 3-4 cm long, diverging at 60° or less; the seed case elongate and longitudinally ribbed. Plate 49, figure 1.

Range: Native of the Prairie provinces, much of the United States and southward to Guatemala, commonly along the banks of streams.

Commonly planted in British Columbia and occasionally escaping, mainly in the interior, as far north as Fort St John and Prince George.

Acer platanoides L.   Norway Maple

Tree to 30 m tall, with greyish, nearly smooth bark. **Leaves** 10-25 cm wide, wider than long; usually rather shallowly 5- to 7-lobed, the lobes with small side lobes, the lobe tips attenuate; leaf glabrous except for a few hairs in the vein axils beneath. Petiole about equal to the width of the leaf blade, with milky juice. **Flowers** in stiff, projecting or erect, not hanging, panicles; the male and female flowers together. **Fruit** glabrous; the wings diverging at 160-180°; the seed-containing base flat and glabrous. Plate 49, figure 5.

Range: A native of Europe, this widely planted species occasionally seeds into the wild, as at Oak Bay, near Victoria and in the Nelson area.

Note: This species is often mistaken for the Sugar Maple (*Acer saccharum* Marshall), especially in eastern Canada.

Acer pseudoplatanus L.   Sycamore Maple

Tree, potentially up to 30 m tall, with coarsely scaly bark. **Leaves** with 5 blunt or rounded lobes but never divided, with crenate to bluntly serrate margins. **Fruit** wings diverging by about 90°.

Range: Native of Europe; often planted; reported as escaping into woodlands in the Vancouver area (Gerald B. Straley, personal communication).

Hippocastanaceae   Horse-chestnut Family

Trees and shrubs with opposite, palmately or pinnately compound leaves. Flowers bilaterally symmetrical, in terminal panicles. Sepals 4 or 5; petals distinct, unequal, clawed. Stamens 5-9, the disc outside the stamens. Ovary superior, 3-chambered, with 2 ovules in each chamber. Style and stigma 1. Fruit a 3-valved, dehiscent capsule, containing 1 or 2 large seeds with large attachment scars.

Aesculus   Horse-chestnut, Buckeye

Deciduous trees and shrubs. Buds large, with several pairs of scales. Terminal bud present on vegetative shoots, absent from flowering branches. Leaves opposite, palmately compound, without stipules. Panicles terminal, ascending; the panicle branches terminating in perfect flowers, with staminate flowers below. Calyx cuplike or tubular. Petals 4 or 5, with long claws.

Aesculus hippocastanum L.   Horse-chestnut

Tree up to 25 m tall, with a short, stout trunk and spreading branches; thick, coarse twigs with large, gummy buds, the conspicuous opposite leaf scars with several bundle scars in a curve. **Leaves** large, 20-50 cm long, with 5-9 (usually 7) obovate to oblanceolate leaflets tapering to bases and rounded and mucronate at apices; crenate-dentate on the margins. **Flowers** 2-3 cm across, the petals mostly white; the upper petals at least with near-basal yellow streaks that later turn pink or red; the panicle falling after the fruit ripens. **Fruit** a globose, green, leathery capsule with scattered short, weak prickles; splitting into 3 valves to release 1 or 2 large, deep brown, nutlike seeds with large, pale attachment scars. Plate 48, figure 1.

Range: Native of southern Europe; commonly planted here and occasionally invading clearings and second-growth forest in the Lower Mainland and on southern Vancouver Island.

Rhamnaceae   Buckthorn Family

Trees and shrubs with simple, alternate or opposite leaves and small, radially symmetrical, bisexual or unisexual flowers commonly in umbel-like clusters; calyx has 4 or 5 lobes, the short hypanthium lined with a fleshy disc; petals 4-5, or absent; stamens inserted on the edge of the disc, opposite the petals; ovary has 2-4 chambers, with usually 1 ovule per chamber. Fruit a capsule or berrylike drupe.

Key to the Genera of *Rhamnaceae*
1a Leaves pinnately veined; petals, if present, short-clawed; fruit a fleshy
    drupe . . . . *Rhamnus*
1b Leaves palmately 3-veined; petals long-clawed; fruit a capsule
    . . . . *Ceanothus*

Rhamnus   Buckthorn

Small trees or shrubs. Buds scaly or naked. Flowers small, greenish, in small umbels or solitary. Fruit a 2- to 4-seeded drupe.

Key to the Species of *Rhamnus*
1a Small tree; winter buds naked; more than 8 pairs of lateral leaf veins;
    petals present . . . . *R. purshiana*
1b Low shrub; winter buds with scales; less than 8 pairs of lateral leaf
    veins; petals absent . . . . *R. alnifolia*

Rhamnus purshiana DC.   Cascara

Small tree up to 12 m tall, with spreading branches and smooth grey bark; twigs with scaleless buds with small, partially expanded leaves at the onset of winter. **Leaves** alternate, sometimes opposite or almost opposite; deciduous when adult, evergreen on young plants; 6-18 cm long, elliptic to obovate, acute or obtuse at apex; serrulate; glabrous above, pubescent beneath on the 10-15 pairs of prominent lateral veins. **Flowers** 8-40 in axillary, pedunculate, umbel-like clusters; petals 5, scarcely longer than the nearly sessile anthers; style very short, not projecting from the hypanthium. **Fruit** widest at apex, 8 mm long, purplish black; seeds 2-3, not furrowed. Plate 50, figure 1. Photo 57.

Range: British Columbia to California and Montana in moist woods and along streams. In British Columbia, along the coast, including Vancouver Island, northward to Bella Coola; also in the southern Columbia Forest Region of the southeastern interior of the province, eastward to Creston and northwestward to Revelstoke, Sicamous and Adams Lake.

Rhamnus alnifolia L'Heritier — Alder Buckthorn / Alderleaf Buckthorn

Shrub to 1.5 m tall. **Leaves** bright green, thin, 5-10 cm long, on petioles 10 mm long; elliptic to ovate, acute, crenate-serrate; mostly glabrous or finely pubescent on the 5-7 pairs of veins. **Flowers** 2-5 in axillary sessile umbels; no petals. **Fruit** subglobose, 6 mm broad, blue-black, with 3 furrowed seeds. Plate 50, figure 2.

Range: In moist woods and swampy places; from the southeastern interior of British Columbia eastward to Quebec and southward to California. In British Columbia, found in the Kootenay region, northward to Columbia Lake and westward to Salmo.

Ceanothus — Snowbrush, Deerbrush

Shrubs with deciduous or evergreen, petioled leaves with 3 strong veins from base and minute stipules; flowers small, white in our species, in showy compact panicles, the branches of which are rather umbel-like, abbreviated racemes. Petals 5, hooded, tapering to narrow claws; style short, stigma 3-lobed. Fruit a 3-valved capsule.

Key to the Species of *Ceanothus*
1a Leaves evergreen, thick, shiny and often sticky; stipules 1-3 mm long, persistent . . . . *C. velutinus*
1b Leaves deciduous, thin, not shiny or sticky; stipules 3-8 mm long, deciduous . . . . *C. sanguineus*

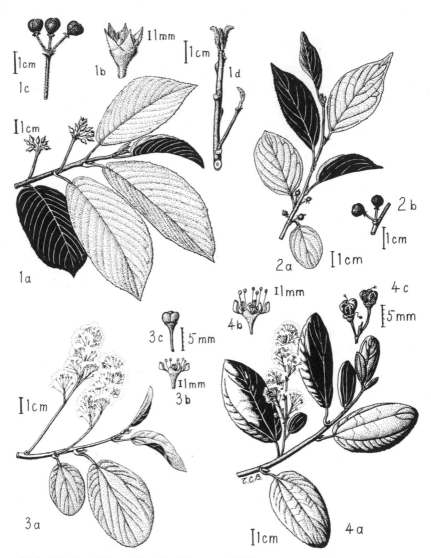

Plate 50 Buckthorn Family. Figures: (1) *Rhamnus purshiana:* 1a, flowering branch; 1b, flower; 1c, fruit; 1d, winter twig; (2) *Rhamnus alnifolia:* 2a, fruiting branch; 2b, fruit; (3) *Ceanothus sanguineus:* 3a, flowering branch; 3b, flower; 3c, fruit; (4) *Ceanothus velutinus:* 4a, flowering branch; 4b, flower; 4c, fruit.

Ceanothus sanguineus Pursh   Redstem Ceanothus

Erect, open shrub up to 3 m tall, with reddish to purplish branches. **Leaves** alternate, broadly elliptic, thin, soft, deciduous, finely crenate-serrulate; usually hairy on the veins beneath; the margins not curved downward. **Flowers** white, in dense panicles of umbels on short lateral branchlets on the previous year's growth. **Fruit** an uncrested 3-valved capsule about 4 mm long. Plate 50, figure 3. Photo 60.

Range: Southern British Columbia to California and eastward to Montana, in woods on gravelly or stony soils. In British Columbia, from the lower Fraser River valley to the West Kootenay region and northward to Little Fort on the North Thompson River and to Cameron Lake on Vancouver Island.

Ceanothus velutinus Douglas ex Hooker   Snowbrush

Rather dense shrub; erect and up to 4 m tall at the coast, but low and spreading and up to 1 m tall, in the interior, where it may form low mounds or be completely prostrate. Branches greenish to greyish. **Leaves** broadly elliptic, thick, leathery, evergreen, with minutely dentate margins; the sides often bent downward, 5-10 cm long; glabrous or hairy beneath, the upper surface commonly shiny; emitting a sweet aroma in warm weather. **Flowers** white, in dense panicles as in *C. sanguineus*. **Fruit** a barely crested, 3-lobed capsule. Plate 50, figure 4. Photo 59.

Key to the Varieties of *C. velutinus*
1a Leaves glabrous beneath; tall, erect shrub; west of Coast and Cascade
   ranges . . . . **var.** *laevigatus* **(Hooker) Torrey & Gray**
1b Leaves finely puberulent beneath; low, spreading shrub; east of Coast
   and Cascade ranges . . . . **var.** *velutinus*

Range: Central British Columbia to California and Colorado, usually on stony or gravelly soils in open woods. In British Columbia, widespread in the interior, northward to Wells Gray Park and westward to Manning Provincial Park and also around Cameron and Horne lakes on Vancouver Island.

Thymelaeaceae   Mezereum Family

Shrubs with tough bark and simple, entire leaves without stipules. Flowers with tubular or funnel-form calyx, often coloured, but no petals. Stamens twice as many as calyx lobes. Ovary 1-chambered, 1-ovuled, free from the calyx. Fruit a drupe.

Daphne   Daphne

Deciduous or evergreen shrub with entire leaves. Flowers perfect; the calyx tubular or funnel-form, with 4 flaring lobes; petals none. Stamens 8; the anthers almost sessile on the inner side of the hypanthium, not projecting out of it. Ovary with a sessile or almost sessile, bead-like stigma. Fruit a drupe.

Daphne laureola L.   Daphne-laurel, Spurge-laurel

Evergreen shrub up to a metre tall, with green branchlets. **Leaves** crowded and oblanceolate, dark green and leathery, commonly curving downward toward their tips. **Flowers** in small axillary clusters, close to the leaf bases, yellowish green and rather inconspicuous; flowering in winter to early spring. **Fruit** a purple or black, poisonous drupe. Plate 52, figure 1.

Range: Native of Europe; cultivated here and escaping into shady woodlands in the Vancouver area, on southern Vancouver Island and on Galiano Island; common locally around Vancouver and Victoria.

Cactaceae   Cactus Family

Trees or shrubs usually with thick, succulent spiny aerial stems and reduced or no leaves. Flowers single, with numerous intergrading sepals and petals, their bases joined into a funnel or tube projecting above the inferior ovary. Stamens numerous, attached to the inner surface of the floral tube. Styles 1, with 3 to many distinct stigmas. Fruit a many-seeded berry or sometimes a capsule.

Opuntia   Prickly-pears

Our only genus. Stems jointed, succulent and green, with alternately and spirally arranged, raised areoles (equivalent to nodes of a conventional stem) bearing tufts of spines, minute sharp bristles and tiny fleshy short-lived leaves. Flowers attached at the areoles of the youngest stem joints. Flower base sunk into the minutely bracted peduncles; the uppermost bractlets intergrading with the outermost sepals. Stigmas 4-8.

Our species are low, cushion-forming shrubs, with the permanent aerial stems prostrate except for the ascending flowering segments and the erect, conspicuous yellow flowers; fruit dry in our species, seldom produced.

Key to the Species of *Opuntia*
1a   Stem joints ellipsoid, not flat, 2-5 cm long; the younger joints easily
     broken off . . . . *O. fragilis*
1b   Stem joints flat, obovate, several times as wide as thick, 5-15 cm long
     when fully grown, firmly attached . . . . *O. polyacantha*

Opuntia fragilis (Nuttall) Haworth   Brittle Prickly-pear

Forming mats or cushions up to 70 cm wide and 20 cm high. Stems easily fragmented; the detached joints readily rooting and forming the main mechanism of reproduction and dispersal here. Spines 2-7 per areole. **Flowers** up to 5 cm wide; the yellow petals obovate, cuneate at base and minutely mucronate at apex. Stamens usually red. Plate 51, figure 1. Photo 63

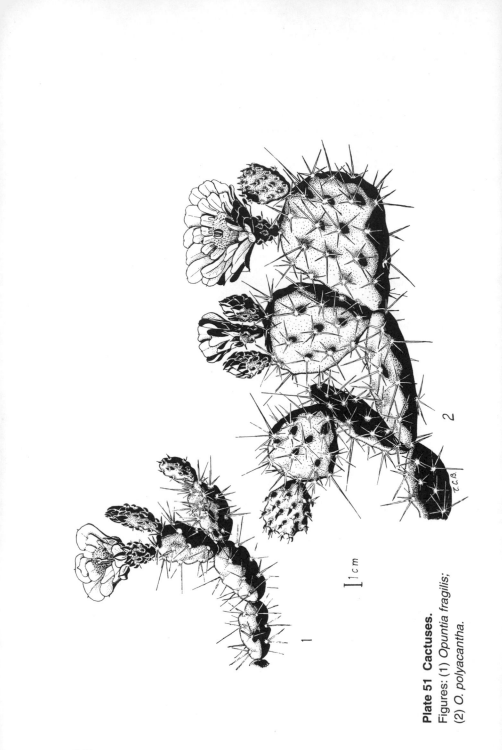

Plate 51 Cactuses.
Figures: (1) *Opuntia fragilis;*
(2) *O. polyacantha.*

Range: British Columbia and Alberta to California and Texas in dry, sunny ground. Widespread in the drier parts of the interior of British Columbia, including the Peace River basin; and on the coast, on rocky bluffs and islands in the Strait of Georgia, northward to Mittlenatch and Savary Islands. This species becomes especially abundant in heavily grazed areas, since it is kicked apart but not eaten by livestock.

Opuntia polyacantha Haworth   Plains Prickly-pear

A coarser plant than *O. fragilis,* forming mats up to 30 cm thick; the firmly attached joints clearly flattened and with 5-11 spines per areole. **Flowers** 5-6 cm wide and high. Petals yellow in our species, and more narrowly cuneate than in *O. fragilis;* stamens yellow or red. Plate 51, figure 2. Photo 61.

Range: Alberta and British Columbia, and southward largely east of the Rocky Mountains to Arizona and Texas. In British Columbia this species is mainly confined to an isolated, narrow belt between Spences Bridge and Maiden Creek in the Bonaparte River valley. It is most abundant near Cache Creek, where it dominates overgrazed pastures. A few records from the south Okanagan Valley area are based on some very scrappy and poorly documented material. This species may have been introduced into its very isolated range in this province at the time of the Cariboo Gold Rush or during the construction of the Canadian Pacific Railway.

Elaeagnaceae   Oleaster Family

Trees or shrubs with branchlets and foliage covered with minute scales that are centrally attached, circular or starlike, and silvery or brown. Leaves alternate or opposite, entire, without stipules. Leaf scars, at least in ours, with 1 bundle scar. Flowers dioecious or perfect, axillary and without petals; hypanthium developed beyond the ovary, persistent, 4-lobed, closely covering the ovary in pistillate flowers or cuplike in staminate flowers. Disc lobed; stamens 4 or 8; ovary superior, 1-loculed and 1-ovuled; style with 1 simple stigma. Fruit drupelike, consisting of the dry stone or nutlet derived from the ovary and the enclosing fleshy hypanthium. Embryo straight with fleshy cotyledons.

Key to the Genera of Elaeagnaceae
1a Leaves alternate; stamens 4 .... *Elaeagnus*
1b Leaves opposite; stamens 8 .... *Shepherdia*

Elaeagnus

Deciduous or evergreen shrubs or trees, often spiny. Winter buds small and ovoid, with few outer scales or none. Leaf scar with 1 bundle scar. Leaves alternate, silvery beneath, with dense stellate hairs. Flowers perfect or polygamous, solitary or in small axillary clusters; hypanthium tubular; calyx 4-lobed; stamens 4, on very short filaments. Fruit contains an ellipsoid striate stone.

Key to the Species of *Elaeagnus*
1a Young twigs brown-scurfy; leaves elliptic, 2-3 times as long as wide
    .... *E. commutata*
1b Young twigs silvery-scurfy; leaves lanceolate, 4 or more times as long
   as wide .... *E. angustifolia*

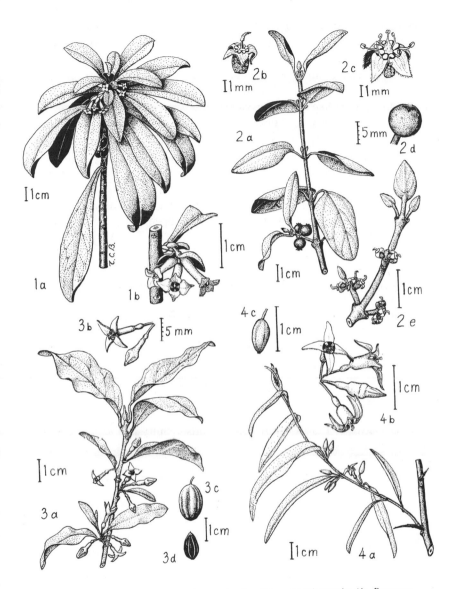

Plate 52 Figures: (1) *Daphne laureola:* 1a, flowering branch; 1b, flowers;
(2) *Shepherdia canadensis:* 2a, fruiting branch; 2b, pistillate flower;
2c, staminate flower; 2d, fruit; 2e, pistillate flowering twig, in March;
(3) *Elaeagnus commutata:* 3a, flowering branch; 3b, flower and bud;
3c, fruit; 3d, fruit stone; (4) *Elaeagnus angustifolia:* 4a, branch with flower
buds; 4b, flowers; 4c, fruit.

Elaeagnus commutata **Bernhardi** *ex* **Rydberg**   **Silverberry**
(*E. argentea* **Pursh not Calla**)   **Wolf-willow**

Erect deciduous shrub up to 5 m tall that suckers freely and forms large colonies; branches reddish brown and unarmed; twigs silvery. **Leaves** 2-10 cm long, ovate to elliptic, cuneate at base, short-petioled; appearing silvery grey from a distance. **Flowers** axillary, 1-3 in a group, silvery outside and yellow within, sweetly fragrant. **Fruit** 1 cm long, ellipsoid, ribbed, silvery, with dry, mealy flesh; the stone longitudinally grooved. Plate 52, figure 3.

Range: Alaska to Quebec and southward to Utah. Widespread in the interior of British Columbia on dry open or semi-open slopes or riverbanks; not on the coast.

Elaeagnus angustifolia **L.**   **Oleaster, Russian-olive**

Tall shrub or small, slender tree; taller than *E. commutata* and less inclined to form dense colonies by suckering. Twigs slender and pliable, usually armed with thorns at the nodes. **Leaves** lanceolate; from a distance the foliage appears whitish rather than silvery. **Flowers** similar to those of *E. commutata*. **Fruit** slenderly ovoid, not ribbed, 7-8 mm long. Plate 52, figure 4.

Range: Native of Europe; introduced in this country and locally escaped in dry open sites near water, as at Penticton, Summerland and Monte Creek.

Shepherdia

Deciduous or evergreen shrubs. Winter buds with 1 or 2 pairs of outer scales, or naked. Leaf scar with 1 bundle scar. Leaves opposite, entire. Flowers dioecious, minute; stamens 8; style slender with 1 inclined stigma. Fruit berrylike or drupelike.

Shepherdia canadensis (L.) Nuttall   Soapberry, Sopoolallie

Erect or spreading, unarmed shrub up to 3 m tall, with brown, scurfy branchlets and buds. Buds flat, ovate, covered with brown-scurfy reduced leaves. **Leaves** 25-50 mm long, ovate, obtuse at base, rounded or obtuse at apex, soft; green above, silvery green beneath and dotted with tiny brown scales. **Flowers** in small clusters, sessile in the axils on bare twigs before leaf emergence, yellowish green; sepals 4; disc prominent, 8-lobed; stamens 8. **Fruit** berrylike, globose to ovoid, scarlet, dotted with pale scales, shiny, at once sweet and astringent. Plate 52, figure 2.

Range: Alaska to the Atlantic coast, and southward to Oregon and the Rocky Mountain states, in open woods and occasionally above the treeline. Common and widespread across British Columbia, including Vancouver Island.

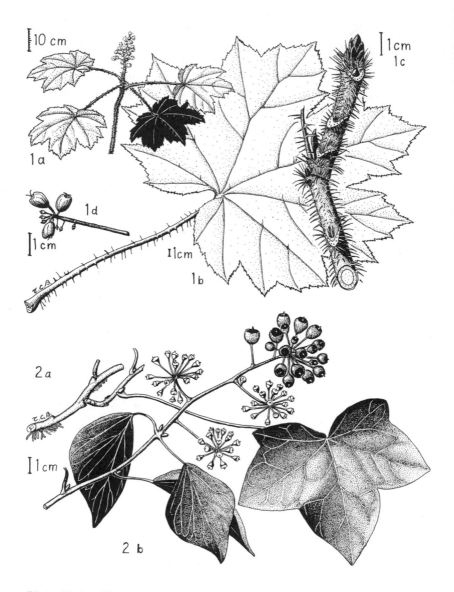

Plate 53 Devil's-club and Ivy. Figures: (1) *Oplopanax horridum:*
1a, top of fruiting plant; 1b, leaf; 1c, winter twig; 1d, fruit; (2) *Hedera helix:*
2a, vegetative, climbing stem with leaf; 2b, fertile branch with fruiting
inflorescence.

254 – ARALIACEAE

1. *Picea glauca* var. *glauca*
(White Spruce)

2. *Picea glauca* var. *albertiana*
(Alberta White Spruce)

3. *Picea glauca* subsp. *engelmannii*
(Engelmann Spruce)

4. *Picea mariana* (Black Spruce)

5. *Tsuga mertensiana*
Mountain Hemlock)

6. *Pseudotsuga menziesii* (Douglas-fir)

7. *Abies grandis* (Grand Fir)

3. *Abies amabilis*
Amabilis Fir)

9. *Abies balsamea* subsp. *lasiocarpa* (Alpine Fir)

10. *Larix lyallii* (Alpine Larch)

1 *Abies balsamea* subsp *lasiocarpa*
Alpine Fir): krummholz form

12. *Pinus ponderosa* (Ponderosa Pine) parkland

II — TREES AND SHRUBS

13. Young Engelmann Spruce
invading old Lodgepole Pine stand.

14. *Pinus banksiana* (Jack Pine)

15. *Pinus flexilis* (Limber Pine)

16. *Juniperus scopulorum*
(Rocky Mountain Juniper)

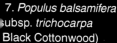

7. *Populus balsamifera*
subsp. *trichocarpa*
Black Cottonwood)

18. *Populus tremuloides*
(Trembling Aspen)

19. *Salix hookeriana*
(Hooker's Willow)

20. *Myrica californica*
California Wax-myrtle)

21. *Alnus crispa*
(Sitka Alder)

22. *Clematis occiden-*
talis (Blue Clematis)

23. *Salix reticulata* (Netted Willow)

24. *Salix arctica* (Arctic Willow)

V — TREES AND SHRUBS

25. *Quercus garryana* (Garry Oak)

26. *Betula glandulosa* (Dwarf Birch)

27. *Polygonum paronychia*
(Black Knotweed)

28. *Eriogonum heracleioides*
(Umbrella Plant)

29. *Berberis aquifolium*
(Tall Oregon-grape)

30. *Clematis ligusticifolia*
(White Clematis)

31. *Berberis nervosa*
(Cascade Oregon-grape)

32. *Philadelphus lewisii*
(Mock-orange)

33. *Holodiscus discolor*
(Ocean-spray)

34. *Physocarpus capitatus* (Pacific Ninebark)

35. *Spiraea densiflora*
(Mountain Spiraea)

36. *Ribes sanguineum*
(Red-flowered Currant)

37. *Dryas octopetala*
(White Dryad) flowers

38. *Spiraea douglasii* (Hardhack)

39. *Spiraea betulifolia* (Flat-topped Spiraea)

40. *Luetkea pectinata* (Partridgefoot)

41. *Dryas drummondii* (Yellow Dryad) fruit

42. *Potentilla fruticosa*
(Shrubby Cinquefoil)

43. *Rubus parviflorus* (Thimbleberry)

44. *Rubus ursinus* (Trailing Blackberry)
staminate flowers

45. *Prunus emarginata* (Bitter Cherry)

46. *Prunus virginiana* (Chokecherry)

47. *Malus diversifolia*
(Pacific Crabapple)

48. *Amelanchier alnifolia*
(Saskatoon Berry)

49. *Sorbus scopulina*
(Western Mountain ash)

50. *Purshia tridentata*
(Bitterbrush)

51. *Oemleria cerasi-formis* (Indian-plum)

52. *Crataegus suksdorfii*
(Black Hawthorn)

53. *Rhus glabra*
(Smooth Sumac)
autumn colour

54. *Paxistima myrsinites* (False-box)

55. *Acer macrophyllum*
(Broadleaf Maple)

56. *Sorbus sitchensis* var. *grayi*
(Sitka Mountain-ash)

57. *Rhamnus purshiana* (Cascara)

58. *Cornus nuttallii* (Pacific Dogwood)

59. *Ceanothus velutinus* (Snowbrush)

60. *Ceanothus sanguineus* (Redstem Ceanothus)

61. *Opuntia polyacantha* (Plains Prickly-pear)

62. *Cladothamnus pyroliflorus* (Copper-bush)

63. *Opuntia fragilis*
(Brittle Prickly-pear)

64. *Oplopanax horridum*
(Devil's Club)

65. *Cornus stolonifera*
(Red Osier Dogwood)

66. *Phyllodoce empet-*
riformis (Pink Heather)

67. *Cassiope tetragona*
(Four-angled
Mountain-heather)

68. *Ledum groenlan-*
dicum (Labrador-tea)

69. *Ledum decumbens*

70. *Kalmia polifolia* (Bog-laurel)

71. *Rhododendron albiflorum*
(White Rhododendron)

72. *Rhododendron macrophyllum*
(California Rhododendron)

73. *Arbutus menziesii* (Arbutus)

74. *Arctostaphylos columbiana*
(Hairy Manzanita)

75. *Vaccinium oxycoccus*
(Bog Cranberry)

76. *Vaccinium ovatum*
(Evergreen Huckleberry)

77. *Phlox longifolia*
(Long-leaved Phlox)

78. *Penstemon fruticosus*
(Shrubby Penstemon)

79. *Vaccinium parvifolium* (Red Huckleberry)

80. *Arctostaphylos uva-ursi* (Bearberry)

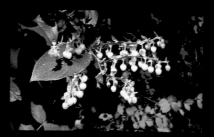

81. *Gaultheria shallon* (Salal)

82. *Fraxinus latifolia* (Oregon Ash)

83. *Phlox caespitosa* (Tufted Phlox)

84. *Linnaea borealis*
(Twinflower)

85. *Viburnum edule*
(Squashberry)

86. *Viburnum trilobum*
(High-bush Cranberry)

87. *Lonicera involucrata*
(Black Twinberry)

88. *Symphoricarpos mollis* (Trailing Snowberry) fruit

89. *Sambucus racemosa*
(Red Elderberry)

90. *Sambucus cerulea*
(Blue Elderberry)

91. *Lonicera ciliosa*
(Orange Honeysuckle)

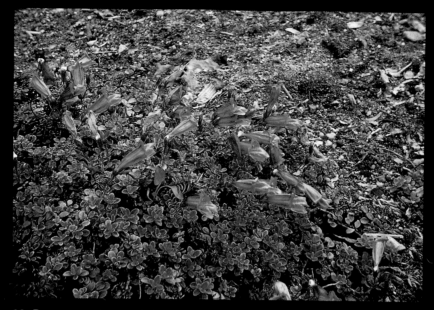

92. *Penstemon davidsonii* (Davidson's Penstemon)

93. *Artemisia tridentata*
(Big Sagebrush) dominating
overgrazed grassland.

94. *Chrysothamnus nauseosus*
(Golden-brush)

XVI – TREES AND SHRUBS

Araliaceae — Ginseng Family

Herbs, shrubs or trees, often prickly; stems usually with large pith. Leaves alternate, entire or lobed, simple or compound. Flowers small, mostly perfect, usually in variously arranged umbels; sepals minute to microscopic; petals usually 5, borne on the edge of the disc, which covers the top of the ovary. Stamens 5; ovary inferior, 1 ovule to each of 2 or more locules. Fruit a drupe or berry.

Key to the Genera of Araliaceae

1a Deciduous, erect prickly shrubs; carpels 2 . . . . *Oplopanax*

1b Evergreen unarmed climber; carpels 5 . . . . *Hedera*

Oplopanax

Deciduous prickly shrubs. Winter buds with several outer scales. Leaves alternate, palmately 5- to 7-lobed, long-petioled. Flowers greenish white, minute, in terminal panicles of umbels. Styles 2, distinct.

Oplopanax horridum (J.E. Smith) Miquel   Devil's Club
(*Fatsia horrida* [J.E. Smith] Bentham & Hooker)

Coarse shrub with ascending, almost unbranched, crooked, thick, yellowish brown, densely prickly stems, 1-4 m high or long, with thick pith; large V-shaped leaf scars each with several bundle scars. **Leaves** orbicular, palmately lobed and veined, 10-45 cm wide, the margin sharply biserrate. Petiole and veins beneath with many prickles; a few prickles on the veins on the upper surface; the blade otherwise glabrous or almost so, thin, bright green. **Flowers** in umbels in conical panicles 10-18 cm long. **Fruit** a rather flat obovoid, scarlet berry, 8-9 mm long. Plate 53, figure 1. Photo 64.
  The slender prickles are breakable and the detached tips can fester when embedded in the skin.

Range: Alaska to Oregon and Montana and around Lake Superior. Widespread across British Columbia, northward to Atlin Lake and the Alsek River valley in the extreme northwest of the province and in the Rocky Mountain foothills west of Fort Nelson; typically in moist soils near streams.

Hedera   Ivy

Evergreen vines or shrubs. Flowers small, greenish and in umbels. Sepals, petals and stamens 5 each; the inferior ovary 5-chambered, but the style single. Fruit a black berry.

Hedera helix L.   English Ivy

Trailing to high-climbing vine, clinging by many short aerial roots. **Leaves** alternate, all along the stems, evergreen, leathery. Leaves have two forms: those on vegetative climbing stems are shallowly, palmately, deltoidly 5-lobed, the blade commonly wider than long; those on non-climbing flowering stems are broadly ovate to rhombic, and entire. **Flowers** in terminal racemes of umbels. **Fruit** black berries in umbels. Flowering and fruiting season often ill-defined; but tending to flower in winter. Plate 53, figure 2.

Range: Introduced from Europe and commonly cultivated. Escaping into woods along the coast from southern Vancouver Island and the adjacent mainland to Oregon. Sometimes aggressively weedy, covering the ground and climbing trees up to 20 or 30 m.

Cornaceae   Dogwood Family

Mostly shrubs or trees, with simple, usually opposite leaves. Flowers are small, 4-parted, borne in heads or open clusters, usually perfect; ovary inferior, usually with 2 chambers containing solitary ovules; styles short. Fruit a drupe or berry, crowned by remains of the perianth. One genus native to British Columbia.

Cornus   Dogwoods

Deciduous trees, shrubs or herbs, with entire leaves opposite in our species; buds elongate and slender with 2 scales meeting edge to edge; the axillary buds appressed to the twig. Leaves entire, the lateral veins curving to become parallel to the margin, usually with appressed hairs. Flowers perfect, in flat cymes or often in heads surrounded by involucral bracts; calyx with minute teeth; petals ovate to oblong; ovary 2-chambered. Fruit a drupe with a 2-chambered stone.

Key to the Woody Species of *Cornus*
1a Flowers in heads with showy white bracts; fruit red; tree
. . . . *C. nuttallii*
1b Flowers in open cymes with vestigial bracts; fruit whitish; coarse
shrub . . . . *C. stolonifera*

Cornus nuttallii Audubon *ex* Torrey & Gray   Pacific Dogwood

Smallish tree with grey smooth bark, becoming scaly in age. Twigs light greyish green, with narrowly tapering buds. **Leaves** 8-12 cm long, ovate to elliptic, obtuse at base, acuminate at apex, thinly appressed-hairy or glabrous, glaucous beneath, with 5-6 pairs of veins; petiole up to 15 mm long. **Inflorescence** a head of minute flowers surrounded by 4-6 large, petal-like white bracts that are obovate to nearly circular, obtuse and 25-50 mm long. **Fruit** crowded on the head, ellipsoid, up to 1 cm long, bright scarlet with a black apical perianth remnant. Plate 54, figure 2. Photo 58.

Range: British Columbia to California and inland to Idaho, in open woods in moist areas. In British Columbia, along the coast to the vicinity of Knight Inlet, to lat. 50°N on the west coast of Vancouver Island, and inland to Manning Provincial Park and Hell's Gate on the Fraser River.

Cornus stolonifera Michaux   Red Osier Dogwood
(*C. sericea* L., in part)   Red-willow

Shrub 2-8 m tall, with ascending stems and commonly arching and rooting branches, tending to spread by suckers and form thickets. Twigs usually deep red. Buds slender and elongate. **Leaves** elliptic, acuminate to acute at tip; the lateral veins curving to become parallel to margins. Underside grey-green and puberulent. **Inflorescence** a flat-topped cyme with puberulent branches and vestigial bracts. **Flowers** white, 5-9 mm across. **Fruit** a white to greyish drupe. Plate 54, figure 3. Photo 65.

Range: A continent-wide species with several local variants. Commonly associated with swamps and alluvial sites.

Key to the Varieties of *C. stolonifera* in British Columbia
1a   Puberulence appressed; petals 2-3 mm long; style 1-2 mm long; stone in fruit smooth; transcontinental shrub, 2-4 m high   .... **var.** *stolonifera*
1b   Puberulence often spreading; petals 3-4 mm long; style 2-3 mm long; stone grooved; taller shrub, up to 8 m tall, from the coast to central B.C.; intergrading with var. *stolonifera*
        .... **var.** *occidentalis* **(Torrey & Gray) C.L. Hitchcock**

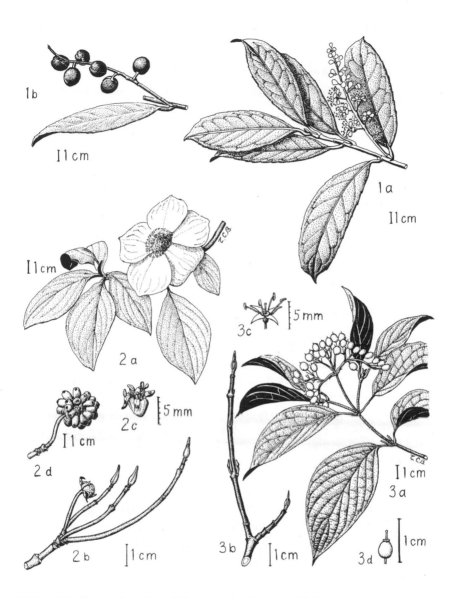

Plate 54 Cherry-laurel and Dogwoods. Figures: (1) *Prunus laurocerasus:* 1a, flowering branch; 1b, fruit; (2) *Cornus nuttallii:* 2a, flowering branch; 2b, winter branch; 2c, flower; 2d, fruiting head; (3) *Cornus stolonifera:* 3a, fruiting branch; 3b, winter branch; 3c, flower; 3d, fruit.

Empetraceae   Crowberry Family

Small evergreen heathlike shrubs. Leaves alternate or roughly whorled, crowded, linear, grooved beneath. Flowers small, with 2 or 3 each of sepals, petals and stamens, and 6-9 coherent carpels with nearly sessile stigmas. Fruit a berrylike drupe.

Empetrum   Crowberry

Flowers of 2 kinds, staminate and pistillate (or, rarely, perfect) appearing on the same or different plants; solitary in the leaf axils; sepals 3; petals 3, reddish, slightly longer than the sepals. Stamens 2-4.

Empetrum nigrum L.   Crowberry

Low, spreading shrub; stems prostrate or ascending to 5 cm (rarely up to 30 cm) and puberulent or glabrous. **Leaves** 4-6 mm long and about a third as wide, spreading and often soon becoming reflexed; thick, glabrous, the margins reflexed until they almost meet, so that the lower surface and midrib are concealed. **Flowers** with 3 sepals 1-1.3 mm long and 3 pink to purplish petals 1.7-2.4 mm long; stamens 4-5 mm long; ovary with a style 0.3 mm long, bearing a 4- to 6-rayed stigma, 1.3 mm wide. **Fruit** black, 6-8 mm in diameter. Plate 55, figure 5.

Range: Widespread in the northern hemisphere, commonly in bogs and other acidic, wet sites from low to alpine elevations; more common northward. In British Columbia, from the coast, including Vancouver Island and the Queen Charlotte Islands, to the Rocky Mountains, and northward and eastward in the Boreal Forest Region.

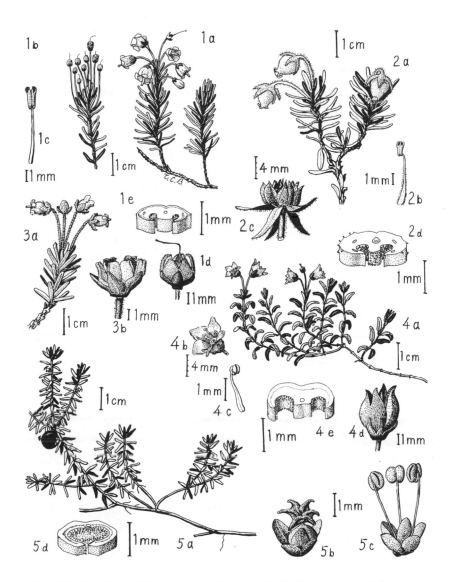

Plate 55 Heaths and Crowberry. Figures: (1) *Phyllodoce empetriformis;*
(2) *P. glanduliflora;* (3) *P. × intermedia;* (4) *Loiseleuria procumbens;*
(5) *Empetrum nigrum.* 1a, 1b, flowering and fruiting branches; 2a, 3a, 4a,
flowering branches; 5a, flowering and fruiting pistillate branch; 1e, 2d, 4e,
5d, cross sections of leaves; 5b, 5c, pistillate and staminate flowers,
respectively; 1c, 2b, 4c, stamens; 1d, 2c, 3b, 4d, fruits (seed capsules).

Ericaceae　　　　　　　　　　　　Heath Family

A large family, mostly of shrubs, but with some trees and some herbs. The leaves are simple, without stipules, usually alternate, and often leathery and evergreen. The flowers are solitary or in racemes, mostly perfect with 4 or 5 petals usually united into a distinctive urn-shaped corolla that faces downward; stamens 5-10, free from the corolla, arising from under the edge of a disc; the anthers opening usually by terminal pores, but sometimes by lateral slits. The anthers are often appendaged in ways distinctive for the species or genus; ovary superior or inferior, usually 5-celled, with a single style and a small stigma. Fruit is a capsule, berry or drupe. For a more detailed treatment, see *The Heather Family of British Columbia* (56).

Members of this family commonly inhabit acidic moist soils and are often found in bogs.

Key to the Woody Genera of Ericaceae

1a　Trees . . . . *Arbutus*
1b　Shrubs.
  2a　Leaves opposite.
    3a　Leaves scalelike, appressed, 3-4 mm long, in 4-angled branch-lets. Flowers axillary.
      4a　Profusely branched shrub up to 1 m tall . . . . *Calluna*
      4b　Sparingly branched, sprawling shrub up to 20 cm tall
      . . . . *Cassiope*
    3b　Leaves spreading, 5-40 mm long. Flowers in terminal groups.
      5a　Leaves 5-8 mm long. Stems more or less prostrate
      . . . . *Loiseleuria*
      5b　Leaves 1-4 cm long. Stems erect . . . . *Kalmia*
  2b　Leaves alternate, although sometimes crowded in apparent whorls.
    6a　Petals distinct or almost so.
      7a　Leaves deciduous. Stems erect. Flowers solitary, terminal on short branches . . . . *Cladothamnus*
      7b　Leaves evergreen.
        8a　Stem creeping and rooting, with ascending branches. Flowers pink.
          9a　Flowers 2 or 3 in a terminal peduncled group. Leaves 2-7 cm long, crowded in whorl-like groups. Petals 5
          . . . . *Chimaphila*

9b   Flowers 1 or 2 on separate peduncles. Leaves 3-10 mm
long, evenly spaced along the threadlike stem. Petals 4
. . . . *Vaccinium*
  8b   Stems erect or decumbent. Flowers many in a short, terminal umbel-like raceme, white . . . . *Ledum*
6b   Petals united into a bell-shaped or urn-shaped corolla.
  10a   Leaves linear to lanceolate; length more than 4 times width.
    11a   Leaves 25-40 mm long, not crowded, glaucous
. . . . *Andromeda*
    11b   Leaves 3-15 mm long, crowded, deep green. Dwarf
shrubs under 20 cm high.
      12a   Leaves 3-5 mm long. Flowers wider than deep
. . . . *Harrimanella*
      12b   Leaves 8-15 mm long. Flowers longer than wide
. . . . *Phyllodoce*
  10b   Leaves relatively broad, length not more than 4 times
width.
    13a   Ovary inferior. Fruit a berry . . . . *Vaccinium*
    13b   Ovary superior.
      14a   Corolla 4-lobed. Flowers not in racemes. Leaves
deciduous . . . . *Menziesia*
      14b   Corolla 5-lobed. Flowers in terminal racemes.
        15a   Fruit a berry or berrylike drupe . . . *Arctostaphylos*
        15b   Fruit a dry, exposed capsule.
          16a   Corolla funnel shaped or bell-shaped, 10-30
mm long . . . . *Rhododendron*
          16b   Corolla tubular or narrowly urn-shaped, 5-6
mm long . . . . *Chamaedaphne*
        15c   Fruit a capsule almost enclosed by a fleshy calyx
. . . . *Gaultheria*

Phyllodoce   **False Heather**

Low, tufted shrub with numerous evergreen alternate linear leaves.
Flowers nodding on erect axillary pedicels, the flower clusters appearing
terminal. Calyx small, 5-parted; stamens 10; anthers awnless, opening by
small apical chinks. Capsules globose, 5-valved, septicidal, many-seeded.

Key to the Species of *Phyllodoce*

1a Corolla cup-shaped, smooth, pink . . . . *P. empetriformis*
1b Corolla flask-shaped, glandular, white or yellowish

. . . . *P. glanduliflora*

Phyllodoce empetriformis (J.E. Smith) D. Don   Pink Heather

Low, closely-branched, tufted shrub. **Leaves** thick, obtuse, grooved beneath by the revolute margins, 8-15 mm long. **Flowers** fragrant; pedicels finely glandular-puberulent; sepals ovate, 2-3 mm long, minutely fringed; corolla cup-shaped, 4-7 mm long. Plate 55, figure 1. Photo 66.

Range: Yukon and southern Alaska to California and Montana. In British Columbia, widespread and abundant in the mountains, but absent from the Queen Charlotte Islands and the northeastern interior; found in open sub-alpine woods and just above the treeline, especially on rocky ground.

Phyllodoce glanduliflora (Hooker) Coville   Yellow Heather
                                                 White Heather

Low, many-branched shrub up to 30 cm tall. **Leaves** obtuse, grooved beneath, 5-9 mm long, sometimes minutely serrulate, glandular. **Flowers** and pedicels glandular-pubescent; sepals lanceolate, acute, 4-6 mm long; corolla flask-shaped, glandular, white or pale yellowish, 6-9 mm long. Plate 55, figure 2.

Range: Yukon and southern Alaska to Oregon and Wyoming. Widespread in British Columbia, including the Queen Charlotte Islands; found on the mountains near and above the treeline, commonly on rocky ground.

Note: *Phyllodoce × intermedia* (Hooker) Camp (*P. empetriformis × glanduliflora*) is the natural hybrid between the two species described above. It is often found in areas where they grow together, and is widely recorded about this province. Plate 55, figure 3.

Loiseleuria Trailing-azalea, Alpine-azalea

Low, spreading shrubs. Leaves small, evergreen, opposite, leathery. Flowers in small terminal groups; calyx 5-lobed, persistent; corolla bell-shaped, 5-lobed; stamens 5, shorter than the corolla; anthers opening by lateral slits. Fruit a 2- or 3-chambered septicidal capsule.

Loiseleuria procumbens (L.) Desvaux Trailing-azalea
Alpine-azalea

Low shrub up to 20 cm tall; branches spreading, sometimes over a metre long. **Leaves** small, 5-8 mm long, narrowly elliptic to lanceolate, margins revolute, midrib prominent beneath. **Flowers** in terminal groups, few, small, white or pink. **Fruit** ovoid. Plate 55, figure 4.

Range: Circumboreal in mountains, at alpine levels on rocky soils. In British Columbia, in the Cassiar Mountains and southward in the Rocky Mountains to the Peace River area and in the Coast Range north of Vancouver; also on Vancouver Island and the Queen Charlotte Islands.

Cassiope Moss-heather, Mountain-heather

Low heatherlike shrubs with numerous small, opposite, 4-ranked overlapping leaves. Flowers solitary, axillary, nodding; corolla bell-shaped, with 4 or 5 lobes; calyx small, with 4 or 5 parts; stamens 8-10; anthers awned, opening by apical pores; style slender; capsule globose and loculicidal.

Key to the Species of *Cassiope*
1a Leaves grooved along the midline beneath .... *C. tetragona*
1b Leaves not grooved beneath.
    2a Branchlet with foliage about 3-4 mm thick. Leaves not, or scarcely, translucent-margined .... *C. mertensiana*
    2b Branchlet with foliage 1-3 mm thick. Leaves translucent-margined
    .... *C. lycopodioides*

Cassiope mertensiana (Bongard) D. Don   Moss-heather
White Mountain-heather

Spreading shrub up to 20 cm tall, with puberulent stems; the shoots, with their leaves, about 4 mm thick. **Leaves** opposite, 4-ranked, appressed to stem, ovate to lanceolate, 3-6 mm long, thick, not grooved dorsally, usually glabrous. **Flowers** on slender axillary pedicels; sepals ovate, obtuse, glabrous, reddish; corolla 5-lobed a quarter to a third of its length, 6-8 mm long, white. **Fruit** 4 mm long. Plate 56, figure 1.

Range: Southern Alaska to Oregon and Montana; throughout British Columbia in the mountains, except in the northeast. Found from the tree-line upward into alpine levels, commonly on rocky terrain.

Cassiope tetragona (L.) D. Don   Four-angled Mountain-heather

Low, spreading shrub with erect branches, up to 20 cm tall; the shoots, with their leaves, 4-6 mm thick. **Leaves** 3-6 mm long, opposite, appressed, grooved dorsally, usually finely short-hairy. **Flowers** few, on glabrous axillary pedicels that at flowering time are 3-4 times as long as the leaves in the northern var. *tetragona,* with corollas 5-7 mm long, but only twice as long as the leaves in var. *saximontana* (Small) C.L. Hitchcock, with corollas 3-5 mm long. Corollas lobed about a third of their length. **Fruit** a capsule similar to that of *C. mertensiana.* Plate 56, figure 2. Photo 67.

Range: A circumboreal species, extending southward through British Columbia to Washington and Montana. Widespread in British Columbia east of the Coast Range, on rocky alpine tundra, mainly as var. *saximontana.*

Cassiope lycopodioides   Club-moss Mountain-heather
(Pallas) D. Don

Evergreen shrub with nearly prostrate branches, up to 20 cm high, but usually less. Branchlets, with foliage, 2-3 mm thick. **Leaves** in 4 sometimes indistinct ranks, closely appressed, rounded and glabrous on the back, 2-4 mm long; the margins ciliate, at least when young and with a translucent border. **Flowers** single, nodding, on axillary pedicels up to 1 cm long but usually only half as long; the corolla 5-8 mm long, bell-shaped, lobed about half to a third of its length. **Fruit** a globose capsule about 3 mm long. Plate 56, figure 3.

Key to the Subspecies of *C. lycopodioides*

1a Leaves 2-3 mm long; translucent border distinct; all cilia very short and straight, apical hairs like the others . . . . **subsp.** *lycopodioides*

1b Leaves up to 4 mm long; translucent border narrow and indistinct; cilia longer; the apical tuft long, prominent, curly, often brownish

. . . . **subsp.** *cristapilosa* **Calder & Taylor**

Range: Subsp. *lycopodioides* is found in alpine situations, from west-central British Columbia to Alaska and eastern Asia; subsp. *cristapilosa* occurs in subalpine and alpine situations on the Queen Charlotte Islands and on the Brooks Peninsula on the west coast of Vancouver Island (9).

Harrimanella

Matted prostrate shrubs with threadlike stems. Leaves numerous, alternate, spreading, short-stalked or almost sessile. Flowers solitary, terminal, turned to one side as in *Cassiope;* corolla lobed at least halfway to base; stamens with stout filaments and awned anthers opening by terminal pores. Style short, bulbous-based. Fruit a loculicidal capsule.

Harrimanella stelleriana (Pallas) Coville   Alaska Moss-heath
(Cassiope stelleriana **[Pallas] DC.)   Alaskan Mountain-heather**

Low mat-forming shrub; the branches threadlike, glabrous or almost so. Vegetative branches prostrate, but flowering branches can rise up to 15 cm, although usually less. **Leaves** oblanceolate, 2-3 mm long, obtuse at apex. **Flowers** on terminal pedicels barely exceeding the leaves; corolla widely bell-shaped, 5-6 mm wide, 5-lobed, pink or white. **Fruit** a globose capsule, which may be surrounded by the persistent withered petals. Plate 56, figure 4.

Range: British Columbia and the Yukon through southern Alaska to Kamchatka and Japan. In British Columbia, in the Cascade, Coast and Cassiar Mountains and on the Queen Charlotte Islands. Found on moist alpine slopes, near or above the treeline in the mountains.

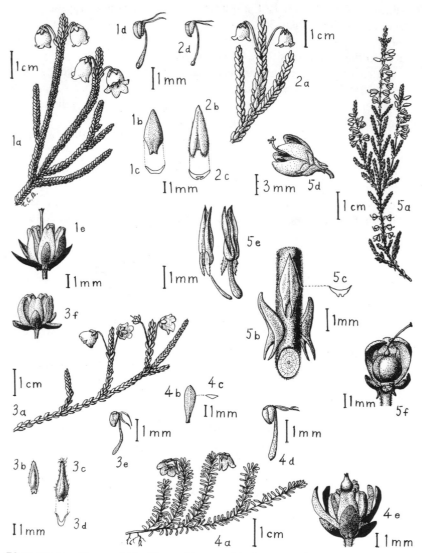

Plate 56 Heaths and Heathers. Figures: (1) *Cassiope mertensiana;* (2) *Cassiope tetragona;* (3) *Cassiope lycopodioides;* (4) *Harrimanella stelleriana;* (5) *Calluna vulgaris.* 1a, 2a, 3a, 4a, 5a, flowering branches; 1b, 2b, 3b, 4b, dorsal views of leaves; 3c, leaf of *Cassiope lycopodioides* subsp. *cristapilosa;* 5b, twig segment with leaves; 1c, 2c, 3d, 4c, 5c, cross sections of leaves; 1d, 2d, 3e, 4d, 5e, stamens; 1e, 3f, 4e, 5f, fruits (seed capsules).

Calluna   Heather, Ling

Spreading to erect shrubs with fine, puberulent twigs and opposite minute leaves. Flowers on short axillary branchlets, with paired bractlets, 4 conspicuous coloured sepals and smaller petals; the petals withering and persistent with the sepals after flowering. Stamens 8, the anthers basally appendaged, opening by longitudinal slits. Fruit a small capsule concealed by the sepals.

Calluna vulgaris (L.) Hull   Heather, Ling

Profusely branched, fine-textured shrub up to a metre tall. Twigs slender, puberulent. **Leaves** opposite, mostly crowded, sessile, minute, each with a pair of basal lobes, glabrous or tomentose. **Flowers** solitary from the leaf axils, each flower with 1-3 pairs of minute, ciliate bractlets, 4 conspicuous, pink or sometimes white sepals, 4 shorter and narrower petals. Stamens with anthers opening by longitudinal slits that do not reach the bases of the anthers, which bear basally attached, downwardly pointing, flat, toothed appendages. Style exserted, slender, with a minute bulbous base and a 4-lobed stigma. **Fruit** a minute, 4-lobed, hairy capsule, concealed under the incurved sepals, only the style exserted. Plate 56, figure 5.

Range: Native of Europe, where it is common on open or wooded, well-drained, acidic soils. A common ornamental garden shrub in this country; escaped locally at Ucluelet, Lulu Island and, rather unaccountably, at Lake Azouzetta. The great variability seen among cultivated varieties may be reflected in naturalized populations when larger collections are made.

Chimaphila   Prince's-pine, Pipsissewa

Low evergreen subshrubs or perennials with creeping, partly woody green stems. Leaves alternate, sometimes crowded and appearing whorled, stiff, serrate. Flowers nodding in terminal clusters, with 5 nearly free sepals, 5 free circular petals, 10 stamens with broad-based, hairy filament bases; the anthers opening by terminal pores. Ovary superior, becoming a 5-chambered, many-seeded, erect, globose, grooved capsule.

Key to the Species of *Chimaphila*

1a Leaves widest beyond mid-length; flowers 4-9 .... *C. umbellata*
1b Leaves widest at mid-length or below; flowers 1-3 ..... *C. menziesii*

Chimaphila umbellata (L.) Barton   Prince's-pine
subsp. *occidentalis* (Rydberg) Hulten

Erect, 10-30 cm tall, with greenish stems. **Leaves** tapering to base, acute-tipped, shiny, indistinctly veined, 2-7 cm long, with widely spaced teeth. **Flowers** 4-9; sepals and petals spreading, pink; filament bases ciliate-edged. **Capsules** grooved, 6-8 mm in diameter. Plate 57, figure 1.

Range: The species is circumboreal; subsp. *occidentalis* is found in western North America, from Alaska to California and Colorado. Widespread in British Columbia, northward in the interior to about lat. 59°N, in coniferous woods at altitudes up to 1,220 m.

Chimaphila menziesii   Menzies' Pipsissewa
(R. Brown) Sprengel

Slender, reddish-stemmed shrub 8-20 cm tall. **Leaves** alternate but close together in threes at the end of each year's growth, closely serrate to entire, sometimes variegated. **Flowers** 1-3; sepals and petals reflexed; the petals whitish to pink; filament bases hairy all around. **Capsules** grooved, 5-7 mm in diameter. Plate 57, figure 2.

Range: British Columbia to California, between the coast and the Cascade and Coast ranges; and in the Columbia Forest Region in Idaho, Montana and British Columbia, in shady coniferous woods. In British Columbia, along the coast from Bella Coola southward, including Vancouver Island; also around Kootenay Lake in the Columbia Forest Region.

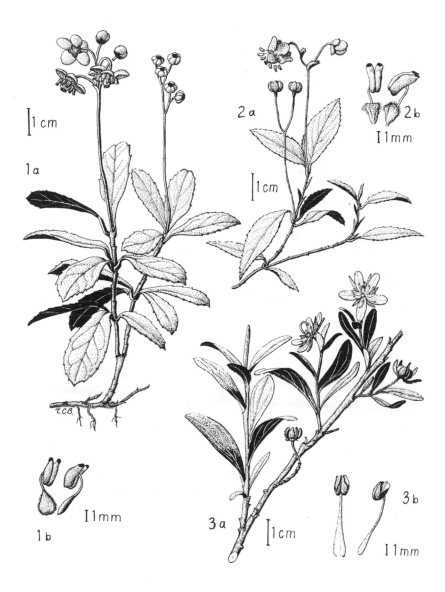

Plate 57 Figures: (1) *Chimaphila umbellata;* (2) *Chimaphila menziesii;* (3) *Cladothamnus pyroliflorus.* a, flowering and fruiting branches; b, stamens.

Cladothamnus

Erect shrub up to 2.5 m tall. Leaves alternate, entire, deciduous. Flowers perfect, usually solitary at ends of branches; sepals 5, nearly or quite distinct; petals 5, distinct; stamens usually 10, anthers basally attached, opening by a slit; ovary superior; style elongated and curved. Fruit a 5- to 6-chambered capsule.

Cladothamnus pyroliflorus Bongard   Copper-bush

Shrub up to 2 m tall, with exfoliating bark. Twigs light brown, minutely puberulent in lines. **Leaves** oblanceolate, tapering to acute; almost sessile at base, blunt or mucronate at tip; entire, light green and rather shiny; glabrous, or minutely puberulent when young. **Flowers** on short terminal puberulent pedicels; sepals oblong-linear, 6-12 mm long; petals widely spreading, oblanceolate, 12-18 mm long, salmon-pink to copper coloured; anthers opening by short apical-lateral slits; style long, upcurved. **Fruit** a depressed-globose, 5-lobed, septicidal capsule, 6-7 mm across, surrounded by the sepals. Plate 57, figure 3. Photo 62.

Range: Alaska to Oregon along the coast. In British Columbia, mostly in the coastal mountains, including Vancouver Island and the Queen Charlotte Islands, in subalpine forest.

Ledum   Labrador-tea

Shrubs with evergreen, alternate, leathery, fragrant leaves tomentose or glandular beneath, often revolute. Flowers white in terminal, umbel-like racemes from scaly buds. Calyx 5-parted; petals distinct, spreading; stamens 5-10; anthers small, basally attached, opening by terminal pores; ovary superior. Fruit a capsule opening from the base into 5 valves.

Key to the Species of *Ledum*
1a   Leaves scaly and glandular beneath, not or scarcely revolute
.... *L. glandulosum*
1b   Leaves densely tomentose beneath; the margins revolute.
   2a   Leaves lanceolate, 2-5 cm long by 5-8 mm wide; flowers 8-14 mm
   wide .... *L. groenlandicum*

2b   Leaves linear, 1-2.5 cm long by 1-2 mm wide; flowers 6-9 mm
wide . . . . *L. decumbens*

Ledum groenlandicum Oeder   Labrador-tea
(*L. palustre* L. subsp. *groenlandicum* [Oeder] Hulten)

Shrub up to 1.5 m tall, with hairy twigs. **Leaves** 25-45 mm long, lanceolate, rounded at base and apex, margins strongly revolute; glabrous and
rather wrinkled above, densely tomentose beneath, the initially pale hairs
soon turning reddish brown, obscuring the surface. **Flowers** white, stamens 5-10. **Fruit** ovoid, 5-6 mm long. Plate 58, figure 2. Photo 68.

Range: Transcontinental, in bogs and acidic swamps. Widespread across
British Columbia.

Ledum decumbens (Aiton) Loddiges *ex* Steudel
(*L. palustre* L. subsp. *decumbens* [Aiton] Hulten)

Low shrub, like a dwarfed form of *L. groenlandicum,* often cushion-forming in exposed situations, with brown-hairy branchlets. **Leaves** linear, 1-
2.5 cm long, strongly revolute-margined; wrinkled above, with reddish
brown, matted hairs beneath. **Flowers** 6-9 mm wide. **Capsules** 3-4 mm
long. Plate 58, figure 3. Photo 69.

Range: Alaska to Greenland at boreal and arctic latitudes, in heaths and
dry, rocky tundra formations. In British Columbia, on the northern mountains north of lat. 58°N, in exposed rocky alpine sites.

Ledum glandulosum Nuttall   Trapper's-tea
Glandular Labrador-tea

Shrub about a metre tall, with minutely puberulent and gland-dotted
branchlets. **Leaves** up to 5 cm long and a third to half as wide; oval;
rounded at base, mucronate to acuminate at apex, the margins scarcely
revolute; glabrous above and not wrinkled, thinly puberulent and glandu-
lar-dotted beneath. **Flowers** in clusters up to 6 cm across, with hairy
pedicels; petals white, about 8 mm long; stamens usually 10, longer than the
short style. **Fruit** a subglobose capsule about 5 mm long. Plate 58, figure 1.

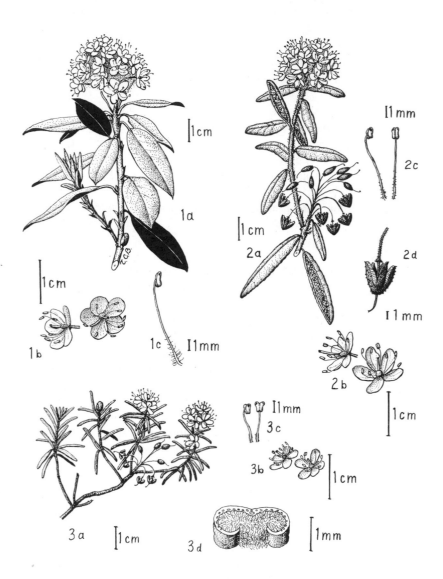

Plate 58 Ledum. Figures: (1) *Ledum glandulosum;* (2) *L. groenlandicum;*
(3) *L. decumbens.* 1a, flowering branch; 2a, 3a, flowering and fruiting
branches; 1b, 2b, 3b, flowers; 1c, 2c, 3c, stamens; 2d, open seed capsule;
3d, cross section of leaf.

Range: British Columbia to Oregon, Idaho and Wyoming, mostly east of the Cascade and Coast ranges, in moist or swampy woods. In British Columbia, from the Cascade Range to the Rocky Mountains, at high montane forest levels or in the interior subalpine forest, northward to Wells Gray Park and westward to Manning Provincial Park and Tazeko Lake.

Rhododendron　　　　　　　　　　**Rhododendrons**

Shrubs with entire alternate petioled leaves without stipules. Flowers showy, in clusters or solitary; calyx 5-lobed; corolla widely funnel-shaped or bell-shaped; stamens 5-10; anthers open by apical pores. Fruit a septicidal capsule that opens from the apex.

Key to the Species of *Rhododendron*
1a　Leaves evergreen, glabrous. Flowers in terminal clusters, pink to purple.
　　2a　Leaves 10-20 cm long, smooth above . . . . *R. macrophyllum*
　　2b　Leaves 0.6-2 cm long, glandular above, scaly beneath
　　. . . . . *R. lapponicum*
1b　Leaves deciduous, hairy; flowers white, axillary . . . . . . *R. albiflorum*

Rhododendron macrophyllum　　　**California Rhododendron**
D. Don *ex* G. Don (*R. californicum* Hooker)

Shrub 1-3 m tall, glabrous. **Leaves** evergreen, 15-20 cm long, elliptic to lanceolate; acute to obtuse at base, acute or mucronate at apex; entire; dark green above, paler beneath. **Flowers** several in a compact terminal raceme; calyx lobes very short; corolla funnel-shaped, slightly bilaterally symmetrical, 3-4 cm across, rose-coloured with darker spots inside the upper lobes; stamens 10, curved, as is the style. **Capsules** 1.5-2 cm long. Plate 59, figure 1. Photo 72.

Range: British Columbia to California in the Cascade and Coast mountains. Generally rare in southern British Columbia, reaching its northern limit in the upper Skagit River valley in the Cascade Mountains; rare on southern Vancouver Island. Found at moderate elevations, commonly rooting in rotting wood on coarse, stony soils. Very showy when flowering in June.

Rhododendron lapponicum (L.) Wahlenberg   Lapland Rosebay

Low, branching shrub, 5-30 cm tall; the branchlets scaly when young. **Leaves** evergreen, oval, 6-20 mm long; glandular above, densely brownish-scaly beneath; on a petiole 2-3 mm long. **Flowers** in terminal clusters of 3-6; corolla funnel-shaped, 10-15 mm across, bright purple; stamens 5-8; ovary covered with tiny circular scales. Plate 59, figure 3.

Range: Circumboreal in alpine heath and arctic tundra, and in open subalpine woods, often on limestones and dolomites. In British Columbia, in the northern Cassiar and Rocky mountains, southward in the Rocky Mountains to about lat. 57°N in British Columbia (and to lat. 52°N in Alberta).

Rhododendron albiflorum Hooker   White Rhododendron

Shrub 1-2 m tall with glandular, hairy twigs. **Leaves** 4-6 cm long by 1.5-2 cm wide, deciduous, narrowly oval, acute at ends, entire, sparsely brown-hairy on both sides. **Flowers** in groups of 1-4, from the axils on the previous year's growth; calyx large, hairy; corolla nearly radially symmetrical, widely bell-shaped, 20-25 mm across, white; stamens with hairy filaments. **Fruit** an ovoid, glandular capsule. Plate 59, figure 2. Photo 71.

Range: British Columbia to Oregon and Montana in open subalpine woods up to the treeline in the mountains. In British Columbia, on the coast, including Vancouver Island and northward at least to Bella Coola, and in the Rocky Mountains northward to lat. 58°N.

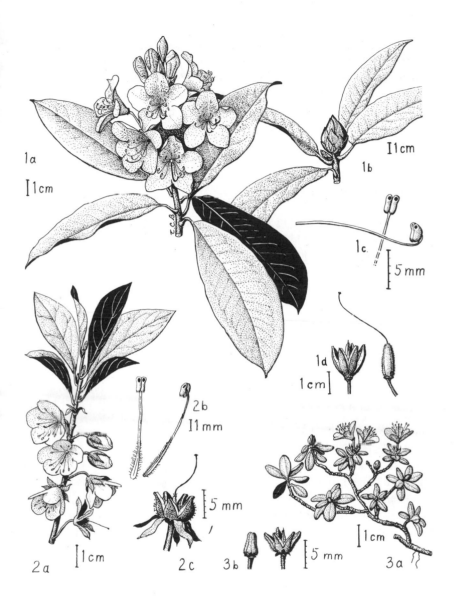

Plate 59 Rhododendrons. Figures: (1) *Rhododendron macrophyllum;*
(2) *R. albiflorum;* (3) *R. lapponicum.* 1a, 2a, 3a, flowering branches;
1b, terminal inflorescence bud; 1c, 2b, stamens; 1d, 2c, 3b, capsules.

Kalmia

Shrubs with glabrous stems. Leaves opposite, very short-petioled, entire, evergreen. Flowers in short, bracted racemes; sepals 5, corolla bowl-shaped, deciduous, pleated to form pouches for the anthers. Stamens 10; filaments elastic, tending to straighten; anthers awnless, held in the pouches in the corolla against the tension of the bent filaments until tripped by insects, opening by apical pores. Ovary superior. Fruit a 5-valved septicidal capsule.

Kalmia polifolia Wangenheim   Bog-laurel

Erect shrub 15-70 cm tall. **Leaves** 1-4 cm long, elliptic to lanceolate; glabrous, dark green and lustrous above, very glaucous beneath; commonly revolute-margined. **Flowers** in terminal leafy-bracted racemes, on pedicels 2-3 cm long; corolla 8-18 mm across, pink to purple. **Fruit** glabrous, globose. Plate 60, figure 1. Photo 70.

Key to the Varieties of *K. polifolia*
1a   Leaves 2-4 cm long, broadly lanceolate to elliptic; corollas 12-18 mm wide; bush 20-70 cm tall .... **var.** *polifolia*
1b   Leaves 1-2 cm long, relatively more broadly elliptic; corollas 8-12 mm wide; bush often only 15-20 cm tall .... **var.** *microphylla*

Range: Boreal-transcontinental, and south in the Cordilleran region to California and Colorado; commonly in Sphagnum bogs. Var. *polifolia* is transcontinental and south mainly in coastal areas to California, generally at low to middle elevations; widespread in British Columbia, including Vancouver Island, but absent from a large area east of the Rocky Mountains. Var. *microphylla* (Hooker) Rehder ranges from Yukon to Colorado via the Rocky Mountains and other interior ranges and to California via the Cascade Range; generally found at subalpine and alpine elevations in wet meadows and by lakeshores.

Menziesia   False-azalea

Deciduous shrubs, with alternate, petioled, entire or minutely serrulate leaves. Flowers in terminal clusters, appearing with the leaves. Stamens 5, 8 or 10. Ovary superior, with 4 or 5 chambers. Capsules 4- or 5-valved, septicidal. Seeds numerous, linear.

Menziesia ferruginea Smith   Rusty Menziesia

Slender shrub up to 2 m (or more) tall, the stems with shredding bark, and the twigs finely puberulent and glandular. **Leaves** elliptic to obovate, up to 6 cm long, cuneate at base, acute and apiculate at apex, minutely serrulate, glandular-puberulent, greyish green; petioles and pedicels glandular-hairy. **Flowers** with parts in fours; calyx small, glandular-margined; corolla ovoid to cylindrical, reddish, 6-12 mm long, 4-lobed; stamens 8, the anthers opening by apical pores. **Fruit** a 4-valved, narrowly ovoid capsule, 5-7 mm long. Plate 60, figure 2.

Range: Alaska to northern California and eastward to Alberta and Wyoming in moist coniferous woods. In British Columbia, in two distinct territories. Var. *ferruginea,* with acute, apiculate-tipped leaves and glandular but otherwise hairless ovaries, ranges from Alaska to Oregon along the coast to the eastern slope of the Coast Range and eastward to Babine Lake, and on Vancouver Island and the Queen Charlotte Islands. Var. *glabella* (Gray) Peck, generally more puberulent but less glandular, its leaves usually obtuse-tipped and ovaries puberulent as well as glandular, ranges through the wet forests on the Rocky Mountains and the Columbia River basin, northwestward to Wells Gray Park and McBride.

Chamaedaphne

Evergreen shrub with alternate, short-petioled, scurfy leaves. Flowers nodding, in terminal racemes; calyx small, with 5 lobes overlapped by a pair of minute bractlets. Fruit a slightly flat, globose, loculicidal capsule.

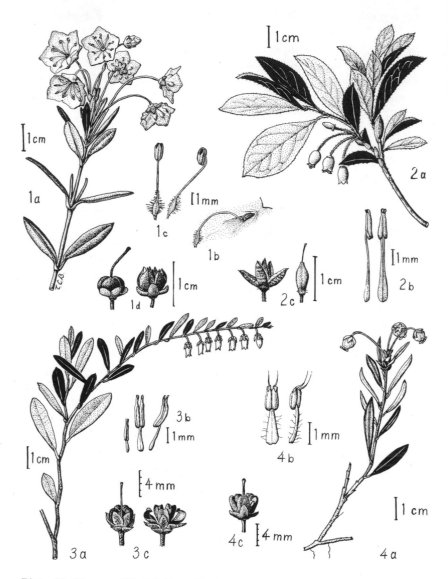

Plate 60 Figures: (1) *Kalmia polifolia;* (2) *Menziesia ferruginea;*
(3) *Chamaedaphne calyculata;* (4) *Andromeda polifolia.*
1a, 2a, 3a, 4a, flowering branches; 1b, stamen in "set" position, anther in
corolla pouch; 1c, stamens sprung and discharged; 2b, 3b, 4b, stamens;
1d, 2c, 3c, 4c, seed capsules.

Chamaedaphne calyculata (L.) Moench   Leatherleaf

Erect shrub up to 1.5 m tall. **Leaves** elliptic to oblanceolate, 10-50 mm long, tapered to base, acute or obtuse at apex, dull olive-green, yellowish-scurfy beneath. **Flowers** in terminal one-sided, bracted racemes, 5-15 cm long, commonly turned to one side; corolla nearly cylindrical, 6 mm long, white. Stamens 10, 5 long and 5 short; anthers tipped by terminal tubular beaks with terminal pores. **Fruit** a capsule 4 mm across, scurfy with minute circular scales; opens by splitting through the seed chambers (locules). Plate 60, figure 3.

Range: Circumboreal at low to moderate elevations, including northeastern British Columbia east of the Rocky Mountains; often a dominant undershrub in swamps and bogs.

Andromeda   Bog-rosemary

Low evergreen shrubs, with alternate entire narrow short-petioled leaves. Flowers small, nodding in short, terminal, umbel-like racemes; calyx 5-lobed; corolla round, urn-shaped, with 5 short recurved lobes; stamens 10, included; anthers awned, opening by apical pores. Fruit a subglobose loculicidal capsule, with numerous oval seeds.

Andromeda polifolia L.   Bog-rosemary

Shrub up to 60 cm tall, often rooting along the stems. **Leaves** 15-40 mm long, lanceolate to linear, acute at ends, revolute-margined, glabrous; bluish green above, white beneath. **Flowers** few on slender pedicels in small terminal racemes; calyx usually reddish; corolla subglobose, pink. **Fruit** a glabrous capsule 5-8 mm across. Plate 60, figure 4.

Range: Circumboreal in bogs. Widespread in British Columbia, including Vancouver Island and the Queen Charlotte Islands; common north of lat. 54°N, but absent from southeastern parts of the province.

Arbutus

Trees or tall shrubs with reddish exfoliating bark. Leaves alternate, evergreen, petioled, entire or serrate. Flowers small in terminal compound racemes; calyx dry, persistent, deeply 5-lobed; corolla urn-shaped, deciduous; stamens 10; anthers awned, opening by slitlike apical pores; ovary superior, 5-chambered; fruit a subglobose, rough-surfaced, several-seeded, berrylike drupe.

Arbutus menziesii Pursh   Arbutus, Madrona

Tree, sometimes reaching a height of 30 m, sometimes with several stems from the base. Bark initially pale greenish and smooth, soon turning reddish, exfoliating in papery layers. **Leaves** 5-13 cm long, elliptic; entire or, on suckers or seedlings, sharply serrate; leathery; lustrous green above, paler beneath; rounded to acute at the ends. **Inflorescence** a large, conical compound raceme, usually ascending to erect at flowering time, drooping in fruit. **Flowers** pendulous, with calyx lobes small, roundish; corolla white, 6-7 mm long, with short lobes reflexed from the narrow orifice. **Fruit** a bright red to orange globose drupe about 1 cm in diameter, with a granular-roughened surface. Plate 61, figure 1. Photo 73.

Range: From British Columbia southward along the coast to Baja California. In British Columbia, in dry, usually rocky open sites along the southern coast, northward to Bute Inlet and Quadra Island and to Tlupana Inlet on the west coast of Vancouver Island; and inland in the lower Fraser River valley to Pitt Lake.

Arctostaphylos

Mostly shrubs, varying from dwarf treelike ones to prostrate mat-forming ones. Bark reddish to dark brown and smooth or flaky. Leaves alternate, petioled or sessile, mostly entire. Flowers small, nodding in downwardly flexed terminal racemes; calyx persistent, 4- to 5-parted; corolla urn-shaped, with 5 short recurved lobes; stamens 8-10, anthers ovoid, with two awns, opening by slitlike apical pores. Fruit a drupe with 4-10 nutlets.

Key to the Species of *Arctostaphylos*

1a   Erect shrub; leaves tomentose, grey-green, 2-6 cm long
. . . . *A. columbiana*
1b   Prostrate shrub; leaves glabrous or almost so, 1-3 cm long.
  2a   Leaves evergreen, entire, thick . . . . *A. uva-ursi*
  2b   Leaves deciduous, crenate, thin . . . . *A. alpina*

Arctostaphylos alpina (**L.**) **Sprengel**   **Alpine Bearberry**

Prostrate shrub with short branches with papery, shredding bark. **Leaves** obovate, crenate-margined, thin; often wrinkled with veins impressed above and raised beneath; deciduous or withering and persistent in skeletonized form; 1-3 cm long. **Flowers** appearing before or with the leaves, in terminal groups of 2-4, about 4 mm long and nearly as wide; the corolla urn-shaped, white to pinkish; anthers with very short awns. **Fruit** a juicy drupe with five 1-seeded nutlets. Plate 61, figure 5.

Key to the Subspecies of *A. alpina*
1a   Leaves ciliate; fruit purple to black . . . . **subsp.** *alpina*
1b   Leaves not ciliate; fruit red **subsp.** *rubra* (**Rehder & Wilson**) **Hulten**

Range: Circumpolar in arctic and alpine tundra. Widespread in the interior of British Columbia northward from about lat. 55°N and southward along the Rocky Mountains to about lat. 51°N.

Arctostaphylos columbiana **Piper**   **Hairy Manzanita**

Erect crooked shrubs up to 3 m tall; bark is dark reddish on older stems and white-tomentose or stiffly hairy on young twigs. **Leaves** 2-6 cm long, broadly ovate, usually acute at apex and rounded at base, greyish green, tomentose; petiole 5-9 mm long. **Flowers** in nodding, hairy compound racemes up to 3 cm long, with small, leafy, lanceolate bracts longer than the pedicels. Corolla 6-7 mm long, white. **Fruit** depressed-globose, orange to brownish, minutely puberulent, 5-8 mm across, dry and mealy but sweet. Plate 61, figure 2. Photo 74.

Range: Southern British Columbia to northern California near the coast. In British Columbia, on southern Vancouver Island and the adjacent mainland coast and islands; northward at least to Tlupana Inlet, Cortes Island

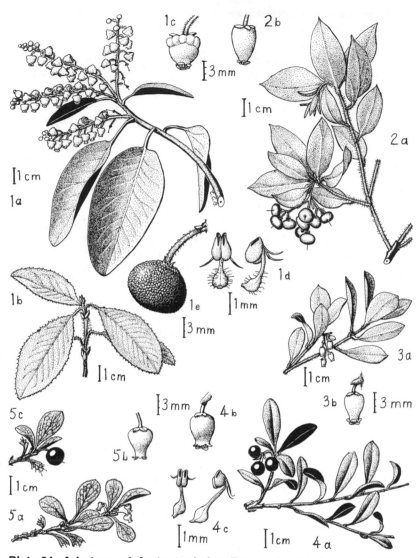

Plate 61 *Arbutus* and *Arctostaphylos.* Figures: (1) *Arbutus menziesii:*
1a, flowering branch; 1b, seedling foliage; 1c, flower; 1d, stamens;
1e, fruit; (2) *Arctostaphylos columbiana:* 2a, fruiting branch; 2b, flower;
(3) *Arctostaphylos × media:* 3a, flowering branch; 3b, flower;
(4) *Arctostaphylos uva-ursi:* 4a, fruiting branch; 4b, flower; 4c, stamens;
(5) *Arctostaphylos alpina:* 5a, flowering branch; 5b, flower; 5c, fruiting
branch.

and Lund. Found on dry rocky slopes, often associated with Kinnikinnick and Arbutus.

Arctostaphylos uva-ursi (L.) Sprengel — Bearberry Kinnikinnick

Low, extensively creeping shrub. Young twigs finely tomentose. **Leaves** obovate to oblanceolate, the blade 15-25 mm long, tapering to a petiole 2-5 mm long; glabrous when mature, deep green above, paler beneath. **Flowers** in short, few-flowered nodding racemes; corolla 4-6 mm long, pink to whitish. **Fruit** globose, 7-10 mm across, glabrous, bright and shiny red, berrylike, not juicy. Plate 61, figure 4. Photo 80.

Range: Circumboreal. Throughout British Columbia on sandy to stony soils in open woods at low to moderate elevations.

Arctostaphylos × *media* Greene

A natural hybrid between *A. columbiana* and *A. uva-ursi*, this is a rather low, broad shrub with a moundlike form, its arching and semi-prostrate branches ascending to 60 cm high; twigs with stiff perpendicular hairs. **Leaves** obovate, narrowed to base and typically obtuse at apex, deep green or paler, slightly pubescent beneath, about 25 mm long or sometimes more. **Flowers** white. **Fruit** red to orange, but seldom produced. Plate 61, figure 3.

Range: Found on Vancouver Island and adjacent islands in company with its parent species.

Gaultheria

Evergreen shrubs. Leaves alternate in ours, short-petioled, usually serrate. Flowers in racemes or solitary; pedicels with 2 bractlets; calyx 5-lobed, increasing in size and becoming fleshy after flowering; corolla urn- or bell-shaped; stamens 8 or 10; ovary glabrous. Fruit a capsule overgrown and enclosed by the fleshy calyx; berrylike.

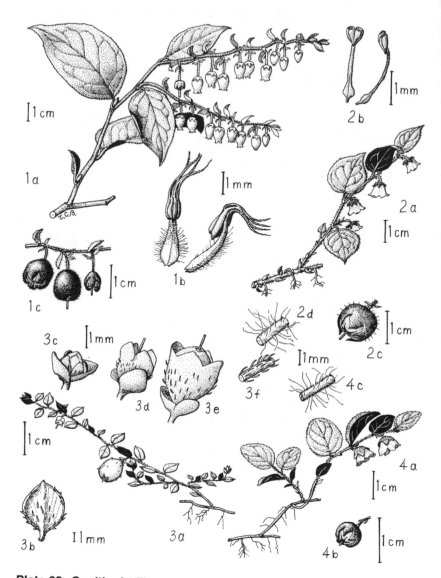

Plate 62 *Gaultheria.* Figures: (1) *Gaultheria shallon:* 1a, flowering branch; 1b, stamens; 1c, mature and half-grown fruit; (2) *G. ovatifolia:* 2a, flowering branch; 2b, stamens; 2c, fruit; 2d, twig vesture; (3) *G. hispidula:* 3a, fruiting shoot; 3b, underside of leaf; 3c, flower bud; 3d, open flower; 3e, developing fruit; 3f, twig vesture; (4) *G. humifusa:* 4a, flowering shoot; 4b, fruit; 4c, twig vesture.

Key to the Species of *Gaultheria*

1a Leaves over 5 cm long; flowers in racemes; fruit purple to black
. . . . *G. shallon*

1b Leaves less than 5 cm long; flowers solitary; fruit red or white.

  2a Stems with stiff, appressed brown hairs; leaves less than 1 cm
  long. Corolla 4-lobed; fruit white . . . . *G. hispidula*

  2b Stems with fine yellowish hairs more or less erect from the surface, or glabrous; leaves commonly more than 1 cm long; corolla 5-
  lobed; fruit red.

    3a Leaves ovate, obtuse at apex; corolla slightly longer than calyx;
    calyx and fruit glabrous . . . . *G. humifusa*

    3b Leaves nearly orbicular, but acute at apex; corolla about twice
    as long as calyx; calyx and fruit pilose . . . . *G. ovatifolia*

Gaultheria hispidula (L.) Muhlenberg   Creeping Snowberry
(*Vaccinium hispidulum* L.; *Chiogenes hispidula* [L.] Torrey &
Gray *ex* Torrey)

Stems slender, trailing, 10-30 cm long, with stiff reclining brownish hairs.
Leaves 4-9 mm long, ovate, acute-tipped; entire or slightly crenate; leathery; shiny and glabrous above, bristly beneath with a few stiff, reclining
hairs; the margins slightly revolute and ciliate. **Flowers** solitary in the leaf
axils, nodding, the base covered by a pair of broad bractlets; calyx and
corolla 4-lobed; stamens 8. **Fruit** white, crowned by the persistent sepal
tips and with scattered short hairs; edible. Plate 62, figure 3.

Range: Transcontinental. Uncommon in British Columbia: in the southeastern and central interior and northwestward to the upper Skeena River
basin; found in mossy woods and muskeg up to subalpine levels.

Gaultheria humifusa (Graham) Rydberg   Alpine Wintergreen
(*G. myrsinites* Hooker)

Closely branched trailing shrub to 5 cm high; the stem sparsely pilose to
almost glabrous. **Leaves** ovate, 5-20 mm long, obtuse or rounded at apex,
entire or nearly so, ciliate. **Flowers** with short pedicels bearing a pair of
bractlets; corolla bell-shaped, slightly longer than the calyx; stamens 10,
similar to those of *G. ovatifolia*. **Fruit** 7-9 mm across, glabrous, red. Plate
62, figure 4.

Range: Subalpine meadows from southern British Columbia to California and Colorado. In British Columbia, from Vancouver Island to the Rocky Mountains; northward in the Coast Range to Mount Waddington and in the Selkirk Range to Glacier National Park.

Gaultheria ovatifolia A. Gray   Western Teaberry

A small, slender shrub up to 30 cm tall, with pilose twigs. **Leaves** orbicular to broadly ovate; acute to short-acuminate at apex, truncate to subcordate at base; stiff and leathery; the margin serrulate and ciliate, at least when young. **Flowers** solitary in the leaf axils; corolla bell-shaped, about twice as long as the calyx; stamens with rough but glabrous filaments and unappendaged anthers opening by apical pores. **Fruit** 7-9 mm across, red, hairy, the top of the capsule visible between the fleshy calyx lobes. Plate 62, figure 2.

Range: Southern British Columbia to California and Idaho, in coniferous forests. In British Columbia, in two areas: on the coast from Vancouver Island to Garibaldi and Manning provincial parks, and in the Selkirk Range northward to Glacier National Park.

Gaultheria shallon Pursh   Salal

Diffuse shrub with stiff, wiry, crooked stems; short, or up to 2 m high. Branchlets glandular-pilose. **Leaves** 5-13 cm long, ovate; acute or acuminate at apex, rounded or subcordate at base; serrulate, stiff and hard and glabrous at maturity. **Flowers** in terminal and axillary racemes; peduncle and pedicels glandular-pilose; bracts ovate to lanceolate, reflexed; each pedicel with a pair of bractlets; calyx lobes triangular, whitish, pubescent; corolla ovoid, white or pink; stamens with stout hairy filaments and anthers opening by short slits at the curve where they taper into a pair of forked appendages. **Fruit** purple becoming black, hairy. Plate 62, figure 1. Photo 81.

Range: From southern Alaska to California, from the coast to the Cascade and Coast ranges. In British Columbia, from the coast inland to Manning Provincial Park and also in the Kootenay Region around Kootenay Lake. In the wetter areas on the coast, it sometimes forms impenetrable thickets.

Vaccinium   Blueberries, Huckleberries, Cranberries, etc.

Erect to prostrate shrubs with slender twigs without terminal buds. Leaves generally small, deciduous or evergreen, entire or serrate. Flowers solitary in leaf axils or in racemes, sometimes apparently terminal; calyx shallowly lobed; corolla urn-shaped, bell-shaped or widely flaring; stamens 8-10; the anthers awned or awnless, opening through prolonged terminal tubes; ovary inferior. Fruit a many-seeded berry, crowned by the persistent calyx lobes.

While Szczawinski (56) treated *Oxycoccus* as a genus separate from *Vaccinium*, the treatment of it here, as a part of *Vaccinium*, follows that of most authors (29, 62 and others). The treatment of its species follows that of Popova (48).

Key to the Species of *Vaccinium*
1a Leaves evergreen.
   2a Stems trailing, with prostrate or ascending branches; petals or corolla lobes 4; fruit red.
      3a Leaves entire, acute at apex, up to 10 mm long, plain white beneath; petals separate almost to base, reflexed.
         4a Leaves elliptic, widest near the middle, 6-10 mm long; pedicels puberulent .... *V. oxycoccus*
         4b Leaves ovate, widest near base, 2-6 mm long; pedicels glabrous or almost so .... *V. microcarpum*
      3b Leaves obscurely serrulate, obtuse to rounded at apex, 10-18 mm long, black-dotted beneath. Corolla deeply cup-shaped, with a 4-lobed rim .... *V. vitis-idaea*
   2b Stems erect; corolla usually 5-lobed.
      5a Up to 2 m tall; twigs usually reddish to brown, cylindrical, puberulent; leaves 1.5-4 cm long; fruit shiny black .... *V. ovatum*
      5b Small, usually under 30 cm tall; twigs fine, green, angled, glabrous; leaves usually 1 cm or less long; usually sterile
         .... **juvenile** *V. parvifolium*
1b Leaves deciduous.
   6a Twigs nearly circular in cross section, fruit black or blue-black.
      7a Leaves entire.
         8a Flowers several to many, in terminal panicles
            .... *V. corymbosum*
         8b Flowers 1-8, terminal or axillary.
            9a Leaves and twigs densely velvety-hairy; flowers in terminal racemes .... *V. myrtilloides*

9b Leaves glabrous; twigs glabrous or slightly pubescent; flowers 1-4 in leaf axils .... *V. uliginosum*
7b Leaves serrulate; flowers single, in leaf axils.
10a Leaves serrulate only toward apex, glaucous beneath; corolla globose .... *V. deliciosum*
10b Leaves crenate-serrulate full length, not glaucous beneath; corolla ovoid .... *V. caespitosum*
6b Twigs angled, particularly when young.
11a Leaves entire, or slightly serrulate at the base.
12a Fruit red .... *V. parvifolium*
12b Fruit black, purple or bluish.
13a Leaves with scattered glands on midvein. Flowers with the leaves. Corolla globose; style exserted ... *V. alaskaense*
13b Leaves glabrous. Flowers before the leaves; corolla ovoid, urn-shaped. Style not exserted .... *V. ovalifolium*
11b Leaves serrate.
14a Twigs pubescent in grooves. Leaves up to 3 cm long. Berries dark red to black .... *V. myrtillus*
14b Twigs glabrous.
15a Twigs ascending, closely branched, green. Leaves up to 1.5 cm long. Low, broomy bush up to 25 cm high. Berry red .... *V. scoparium*
15b Twigs more sparsely branched, spreading, yellowish to reddish. Leaves 2.5-6 cm long. Taller bushes up to 2 m tall. Berry purple to black, without a bloom.
16a Leaves elliptic, blunt to rounded at tip. Flowers globose. Berry dark purple .... *V. globulare*
16b Leaves ovate, acuminate at tip. Flower ovoid. Berry black .... *V. membranaceum*

Vaccinium oxycoccus L.   Bog Cranberry
(*Oxycoccus quadripetalus* Gilibert; *O. palustris* Persoon)

Creeping shrub similar to *V. microcarpum,* but coarser, with brown to black stems. **Leaves** ovate to elliptic, widest toward the middle of the leaf, 4-10 mm long, rounded at base, acute at apex, glabrous, glaucous beneath, the margin revolute; petiole up to 1 mm long, puberulent. **Flowers** 1-3 on terminal pedicels, similar to those of *V. microcarpum,* but the pedicels puberulent, 2-4 cm long; petals 4-7 mm long, pink or white; stamens 3.5-4 mm long, the anthers awnless; style 5-5.5 mm long. **Fruit** dark red, 8-12 mm in diameter. Plate 63, figure 4. Photo 75.

Range: Circumboreal at mid-latitudes; southward to Oregon and Idaho. In British Columbia, from the Peace River southward and across the mainland and on Vancouver Island and the Queen Charlotte Islands. Found mostly in Sphagnum bogs.

Vaccinium microcarpum **Dwarf Cranberry**
(**Turczaninow** *ex* **Ruprecht**) **Schmalhausen** **Small Cranberry**
(*Vaccinium oxycoccus* **L., in part;**
Oxycoccus microcarpus **Turczaninow** *ex* **Ruprecht**)

Creeping shrub with threadlike, reddish brown stems and prostrate or ascending branches. **Leaves** ovate, broadest near base, 3-6 mm long, glaucous beneath, rounded to truncate at base, acute at apex, the margin revolute, ciliate near base, glabrous on the surface; petiole puberulent, 1 mm long or less. **Flowers** 1 or 2 on terminal, erect, glabrous pedicels 17-30 mm long; each pedicel with a pair of bractlets part-way up; calyx 4-lobed, inconspicuous; petals 4, free except at their bases, 4-5 mm long, ovate to lanceolate, recurved, pink or white; stamens 8, 3-3.5 mm long; the anthers awnless, tapering into tubular tips; style 4-5 mm long, projecting beyond the stamens. **Fruit** a red berry 5-7 mm in diameter. Plate 63, figure 3.

Range: Circumboreal to circumpolar. In western North America, from Alaska and Northwest Territories southward to northern Alberta and British Columbia. In British Columbia, on the Queen Charlotte Islands and across the mainland from the latitude of Prince George northward. Found mostly in Sphagnum bogs.

Vaccinium ovatum **Pursh** **Evergreen Huckleberry**

Erect evergreen shrub, 1-3 m tall; twigs hairy, very leafy. **Leaves** up to 4 cm long, persistent, thick, ovate to lanceolate, acute to acuminate at apex, serrate, glossy. **Flowers** in axillary clusters; corolla pink or white; filaments flat, hairy; anthers awnless. **Fruit** blackish and shiny or purplish and glaucous, 4-7 mm across. Plate 63, figure 1. Photo 76.

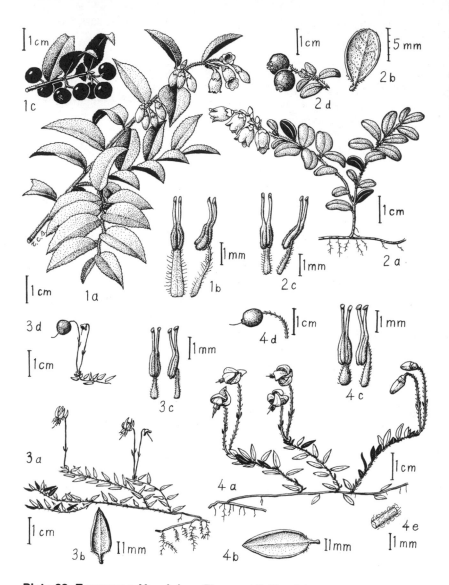

Plate 63 Evergreen *Vaccinium*. Figures: (1) *Vaccinium ovatum;*
(2) *V. vitis-idaea;* (3) *V. microcarpum;* (4) *V. oxycoccus.*
1a, 2a, 3a, 4a, flowering branches; 2b, 3b, 4b, undersides of leaves;
1b, 2c, 3c, 4c, stamens; 1c, 2d, 3d, 4d, fruit; 4e, twig vesture (to compare
with *Gaultheria hispidula,* Plate 62, figure 3f).

292 – ERICACEAE

Range: British Columbia to California along the coast. In British Columbia, on Vancouver Island and the adjacent coast and northward to Port Essington (lat. 54°10'N, long. 129°58'W). Commonly found in semi-open coniferous woods, rooting in decaying litter on stony or rocky soil.

Vaccinium vitis-idaea L.   Red Whortleberry
subsp. *minus* (Loddiges) Hulten   Mountain Cranberry
.   Cowberry, Lingonberry

Low shrub up to 20 cm tall, often matted. **Leaves** usually less than 18 mm long, evergreen, elliptic to obovate, obtuse or rounded at apex, revolute, obscurely serrulate, shiny deep green above, paler beneath and black-dotted. **Flowers** in short terminal clusters; corolla pink, ovoid, 4-lobed; anthers awnless. **Fruit** dark red. Plate 63, figure 2.

Range: Circumboreal in alpine areas. In northwestern and northeastern British Columbia and southward along the Rocky Mountains at least to Yoho National Park and along the coast to the Queen Charlotte Islands and Vancouver Island. Subsp. *vitis-idaea* is in Europe.

Vaccinium myrtilloides Michaux   Velvet-leaf Blueberry

Low shrub, up to 40 cm tall; branches cylindrical in cross section, densely velvety-hairy. **Leaves** elliptic to lanceolate, acute to rounded at base, acute and minutely apiculate at apex, entire, thin, with soft velvety hairs. **Flowers** in short racemes of 5-8 flowers on the ends of the branches, on pedicels shorter than the corollas; flowering while the leaves are still expanding; corollas white and tinged with pink, cylindrical, 5 mm long; filaments hairy; anthers awnless. **Fruit** blue, bloomy berry, 6-10 mm in diameter. Plate 64, figure 1.

Range: From southern British Columbia eastward to the Atlantic coast, northward into the Northwest Territories and southward in the east to West Virginia. In British Columbia, mainly in the interior, northward to the Peace River and westward to Vanderhoof, but with a few records from the Lower Mainland and Vancouver Island.

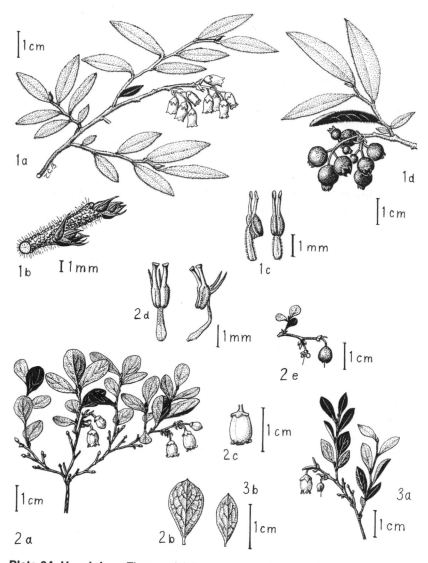

Plate 64 Vaccinium. Figures: (1) *Vaccinium myrtilloides;* (2) *V. uliginosum* subsp. *alpinum;* (3) *V. uliginosum* subsp. *occidentale.* 1a, 2a, 3a, flowering branches; 1b, twig and buds; 1c, 2d, stamens; 2b, 3b, undersides of leaves; 2c, flower; 1d, 2e, fruiting branches.

Vaccinium uliginosum L. **Bog Blueberry**
Alpine Bilberry, Bog Whortleberry

Shrub up to 50 cm tall. Twigs smoothly cylindrical, not angled; glabrous or finely puberulent. **Leaves** narrowly to broadly obovate, 1-3 cm long, cuneate at base and obtuse or rounded at apex, entire, glabrous, often conspicuously veiny beneath. **Flowers** 1-4 in the leaf axils, drooping, about 6 mm long, the floral parts in fours or fives; corolla urn-shaped, pink; anthers awned and tapering into a long tubular beak. **Fruit** a blue-black bloomy berry, 4-8 mm in diameter. Plate 64, figures 2 and 3.

Key to the Subspecies of *V. uliginosum*
1a Leaves usually half or more as wide as long. Leaves strongly veined beneath. Fruit 6-8 mm across .... **subsp.** *alpinum*
1b Leaves usually less than half as wide as long. Leaves not so prominently veined beneath. Fruit 4-6 mm across .... **subsp.** *occidentale*

Range: Subsp. *alpinum* (Bigelow) Hulten: circumpolar in heaths and bogs; in British Columbia, widespread north of the Peace River and southward along the coast and Coast Mountains, including Vancouver Island and the Queen Charlotte Islands. Subsp. *occidentale* (Gray) Hulten is found from the Rocky Mountains to the Pacific coast in the United States. In British Columbia, it is found on central and northern Vancouver Island. Subsp. *uliginosum* is in Eurasia (30).

Vaccinium parvifolium **J.E. Smith in Rees** **Red Huckleberry**

Shrub up to 3 m tall with spreading branches that often form flat sprays; twigs distinctly angled, green, glabrous. **Leaves** on adult bushes usually less than 20 mm long, broadly ovate, round or blunt at base and apex, dull green, entire, late deciduous; leaves on juvenile bushes smaller, 5-15 mm long, ovate, obtuse to apiculate at apex, serrate, evergreen. **Flowers** solitary, axillary, appearing when leaves are half-grown; calyx obscurely 5-lobed; corolla more or less urn-shaped, 4-6 mm long, reddish or greenish; anthers short-awned; style equalling or slightly longer than corolla. **Fruit** globose, 7-10 mm across, red. Plate 65, figure 1. Photo 79.

Range: Alaska to California along the coast and locally inland; in forest openings at low altitudes; often abundant in logged areas, tending to root in old, decaying stumps and logs. In British Columbia, from the coast inland to the Coast and Cascade ranges to Manning Provincial Park; and also at Kootenay Lake and Revelstoke.

Vaccinium ovalifolium J.E. Smith   Oval-leaved Blueberry
Blue Huckleberry

Erect shrub, a metre or more high; twigs conspicuously angled, often reddish. **Leaves** deciduous, 1-5 cm long, ovate to elliptic, rounded to obtuse at base and apex; slightly glaucous beneath, glabrous, not glandular, with prominent veins; the margins entire or sometimes obscurely serrulate near the base. **Flowers** axillary, solitary, appearing with the unfolding leaves, on pedicels 5-8 mm long; calyx almost unlobed; corolla pinkish, ovoid, usually longer than wide, 4-7 mm long; filaments glabrous, shorter than the anthers; anthers with awns a little longer than the tubular apical orifices; style usually shorter than the corolla. **Fruit** on recurved pedicels, globose, up to 1 cm across; purplish black, appearing bluish with a bloom. Plate 65, figure 2.

Range: From Alaska to Oregon along the coast, and locally inland. In the coast forest and ascending to subalpine levels; also in the moist forests of the Columbia Forest Region; associated with *V. parvifolium* at low altitudes.

Vaccinium deliciosum Piper   Mountain Huckleberry
Dwarf Bilberry
Blue-leaved Huckleberry

Low, tufted shrub to 60 cm tall; twigs greenish-brown, glabrous or finely puberulent, inconspicuously angled. **Leaves** thickish, more obovate than oblanceolate or oval, 2-4 cm long; distantly toothed except near the base; usually obtuse or rounded at apex, tapered to base; glaucous beneath when fresh. **Flowers** solitary in axils; corolla pinkish, subglobose, 6 mm long; filaments shorter than anthers, glabrous. **Fruit** 6-8 mm across, deep blue or black with a bloom, nearly globose, sweet. Plate 65, figure 3.

Range: Southern British Columbia to Oregon in the Coast and Cascade ranges, in subalpine meadows. In British Columbia, north to the Forbidden Plateau on Vancouver Island; on the mainland, north to Garibaldi and inland to Manning provincial parks.

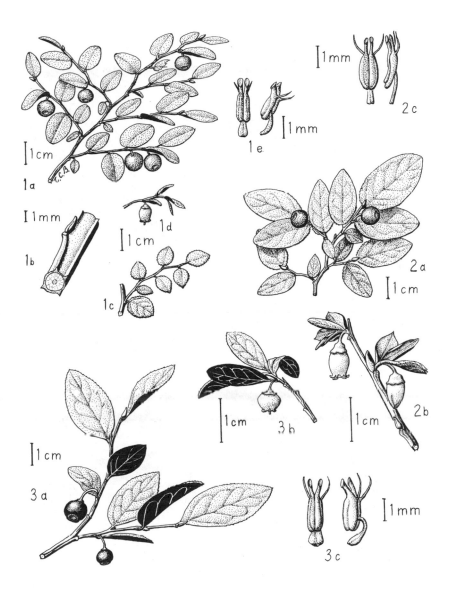

Plate 65 Vaccinium. Figures: (1) *Vaccinium parvifolium;* (2) *V. ovalifolium;*
(3) *V. deliciosum.* 1a, 2a, 3a, fruiting branches; 1b, twig segment;
1c, juvenile foliage; 1d, 2b, 3b, flowering branches; 1e, 2c, 3c, stamens.

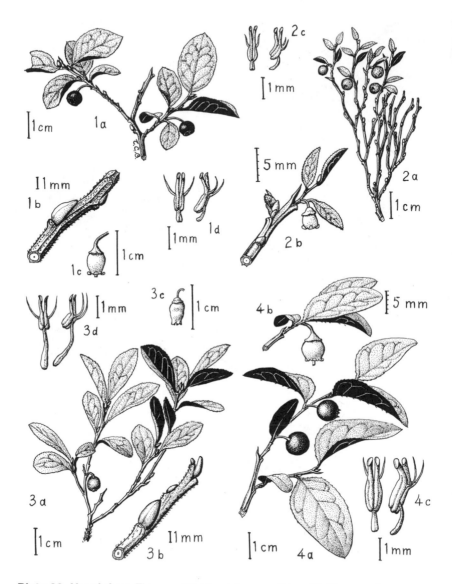

Plate 66 Vaccinium. Figures: (1) *Vaccinium myrtillus;* (2) *V. scoparium;* (3) *V. caespitosum;* (4) *V. alaskaense.* 1b, 3b, twigs and buds; 2b, 4b, flowering branchlets; 1c, 3c, flowers; 1d, 2c, 3d, 4c, stamens.

Vaccinium myrtillus L.   **Low Bilberry**

Low shrub up to 30 cm tall with spreading branches; the angular twigs finely pubescent, mostly between the angles. **Leaves** elliptic to ovate, 1-3 cm long, finely and sharply serrulate, acute or obtuse at apex, glabrous or nearly so, strongly veiny beneath. **Flowers** single, nodding from leaf axils; calyx scarcely lobed; corolla ovoid, pinkish, 5-lobed; stamens with glabrous filaments and awned anthers. **Fruit** dark red to blue-black, without a bloom. Plate 66, figure 1.

Range: British Columbia and Alberta southward to New Mexico; also in Eurasia. In southeastern British Columbia and northward to the Peace River basin, and in the Rocky Mountains; generally at subalpine or alpine altitudes.

Vaccinium scoparium **Leiberg**   **Grouseberry**

Low shrub 12-20 cm tall; twigs numerous, slender and closely branched, erect or ascending, light green, usually glabrous, and sharply angled. **Leaves** 5-13 mm long, ovate-lanceolate to ovate; acute or obtuse at both ends; finely serrate; usually glabrous; conspicuously veined beneath; and nearly sessile. **Flowers** solitary; calyx slightly lobed; corolla pink, bell- or urn-shaped, 2-4 mm long; anthers awned, filaments glabrous. **Fruit** bright red, 3-5 mm across, globose and sweetish. Plate 66, figure 2.

Range: British Columbia and Alberta to California and Colorado. In British Columbia, in the southern interior, westward to the Fraser River and northward in the Rocky Mountains to the Mount Robson area; also at scattered locations in the Coast Range and on the Chilcotin Plateau, between Lillooet and the Fawnie Range. Mainly at subalpine to alpine levels.

Vaccinium caespitosum **Michaux**   **Dwarf Bilberry**
                                                           Dwarf Blueberry

Low shrub up to 30 cm tall, with yellowish green to reddish branches. **Leaves** obovate to oblanceolate, or spatulate, 10-25 mm long, acute or obtuse at apex, more or less crenate-serrulate, thin, green both sides, somewhat glandular beneath. **Flowers** nodding in the axils; calyx with 5 broad, short lobes; corolla ovoid, about twice as long as wide, white or pink, 5-lobed; stamens with glabrous filaments longer than anthers; anthers long-awned. **Fruit** subglobose, 6-8 mm across, bluish to black with a bloom. Plate 66, figure 3.

Range: Alaska to Newfoundland and southward to northern California, in moist alpine meadows and tundra; rarely down to sea level in grassland, as at Victoria. Throughout British Columbia except in the northeastern corner (52).

Vaccinium alaskaense Howell   Alaskan Blueberry
(*V. oblatum* Henry)

Erect shrub up to 1.5 m tall; twigs somewhat angled. **Leaves** ovate to elliptic, up to 6 cm long, with margins entire or inconspicuously serrate on the lower half; sparsely glandular with minute peglike glands on the midvein beneath and commonly slightly hairy beneath. **Flowers** axillary and solitary, on reflexed pedicels 5-15 mm long and somewhat enlarged beneath the calyx; the flowers open when the leaves are about half-grown or more; calyx almost unlobed; corolla depressed-globose, 4-6 mm long, widest at the middle or below; stamens similar to those of *V. ovalifolium;* style usually slightly longer than the corolla. **Fruit** 7-10 mm across, purplish black, with little or no bloom. Plate 66, figure 4.

Range: From Alaska to Oregon along the coast and through the Coast and Cascade ranges. In British Columbia, on Vancouver Island and the Queen Charlotte Islands and inland as far as Babine Lake; from sea level to subalpine levels.

Note: Closely related to *V. ovalifolium.*

Vaccinium membranaceum   Black Huckleberry
Douglas *ex* Hooker

Shrubs 1-3 m tall; young twigs angled, glabrous or slightly puberulent. **Leaves** 25-50 mm long, oval or ovate; apex acute to acuminate, base rounded to obtuse or acute; the margins serrulate for their full length; bright green above, paler green beneath. **Flowers** single, in the axils of the lowest leaves of the current year's growth, on pedicels 5-10 mm long;

calyx shallowly and sinuately 5-lobed; corolla ovoid to urn-shaped, 5 mm long by 4 mm wide, pale yellowish pink; filaments glabrous, shorter than anthers; anthers with awns longer than the tubular beaks; style shortly exserted. **Fruit** 8-10 mm across, purple to black, without a bloom. Plate 67, figure 2.

Range: British Columbia and the southwestern Northwest Territories to Alberta, and southward to Wyoming, northern Utah and northern California, with isolated areas in northern Michigan and Arizona. In British Columbia, on mountain slopes, commonly at subalpine levels; throughout the province except for the extreme northwestern mainland and the Queen Charlotte Islands.

Vaccinium globulare Rydberg

Shrub up to 1.5 m tall; twigs greenish yellow when fresh, glabrous or almost so, and angled. **Leaves** orbicular to obovate or elliptic, obtuse to rounded at apex, 2-5 cm long, serrulate for their full length. **Flowers** single in lower leaf axils; calyx shallowly lobed; corolla more or less globular, varying to ovoid; anthers longer than filaments, tubular-beaked and awned; style equalling the corolla or shortly exserted. **Fruit** a dark purple berry without a bloom. Plate 67, figure 3.

Range: From British Columbia and Alberta southward to Idaho, Utah and Wyoming. In British Columbia, from Kootenay Lake eastward to the Rocky Mountains and northward to Yoho National Park. Found in coniferous forests at low altitudes.

Note: Weakly distinguished from *V. membranaceum*, *V. globulare* is treated by Vander Kloet (58) as synonymous with the former species. Since it is recognizable, this shrub might better be treated as a subspecies of *V. membranaceum*, but this combination has not yet been published.

Vaccinium corymbosum L.   Highbush Blueberry

Erect shrub up to 4 m tall; branchlets cylindrical or the youngest twigs with low, rounded angles, minutely puberulent. **Leaves** elliptic, varying to lanceolate, ovate or obovate; 4-8 cm long; acute at base, acute or acuminate and minutely apiculate at apex; generally entire, sometimes slightly revolute; puberulent on the veins above and pale green and glabrous beneath, where the veins are prominent. **Flowers** in a terminal compound raceme, the branch racemes arising on the leafless end of the stem from buds distinct from those that give rise to leafy branchlets; calyx with 5 acute lobes that are persistent on the fruit; corolla cylindrical to ovoid, 6-12 mm long, white or sometimes pinkish; stamens 10, with hairy filaments and awnless anthers with long tubular beaks. **Fruit** a blue-black, bloomy berry, 6-12 mm across, crowned by the calyx. Plate 67, figure 1.

Range: Native to the eastern United States, and in Canada from Nova Scotia to Quebec and southern Ontario, in swamps and moist woods. In British Columbia, it has long been cultivated in peat-lands of the Fraser River delta for commercial blueberries; and escaped locally in Vancouver and Burnaby and on Lulu Island in open peaty sites.

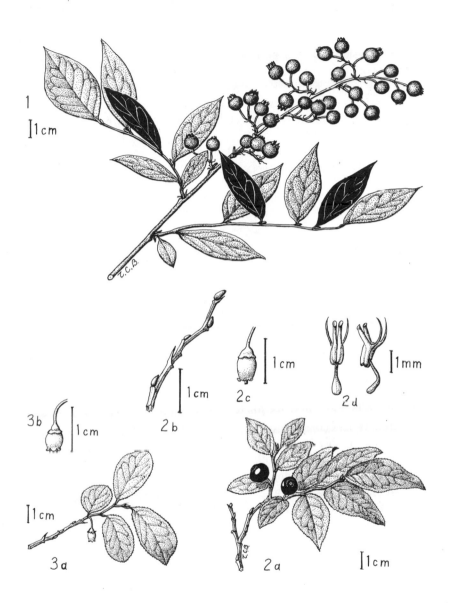

Plate 67 *Vaccinium.* Figures: (1) *Vaccinium corymbosum;*
(2) *V. membranaceum;* (3) *V. globulare.* 2a, fruiting branch; 2b, twig and
buds; 2c, 3b, flowers; 2d, stamens.

Oleaceae   Olive Family

Trees and shrubs with opposite leaves; flowers with 4 sepals, a corolla of 4 petals or none, 2 stamens and a superior 2-chambered ovary.

Key to the Genera of Oleaceae in British Columbia
1a   Leaves simple; flowers perfect; petals present and coherent; fruit a shiny black berry . . . . *Ligustrum*
1b   Leaves pinnately compound; flowers dioecious; petals absent. Fruit a terminally winged samara . . . . *Fraxinus*

Ligustrum   Privet

Shrubs with slender twigs; leaves opposite, simple, entire; flowers in terminal panicles; the 4 petals united along their basal halves into a short tubular or funnel-shaped corolla. Two stamens attached to the corolla; fruit a shiny black berry with one or two seeds.

Ligustrum vulgare L.   Privet

Shrub up to 2 m tall; twigs finely puberulent or glabrous; **Leaves** elliptic to broadly lanceolate, 3-6 cm long by 1-2 cm wide, acute, entire and glabrous, variably evergreen. **Flowers** in small dense panicles, with cuplike, unlobed calyces; corolla white, 4-6 mm wide, the corolla tube as long as the free, diverging lobes; anthers shorter than the corolla lobes; **Fruit** black, shiny and firm, 7-9 mm in diameter, with 1 or 2 seeds, poisonous. Plate 68, figure 3.

Range: A native of Europe, introduced here as a popular subject for hedges and escaping locally between Vancouver and White Rock on the mainland, on southern Vancouver Island, and at Sandspit on the Queen Charlotte Islands.

Fraxinus   Ashes

Mostly deciduous trees with grey bark, stout twigs, and dark, hairy, pyramidal terminal buds; opposite, pinnately compound leaves; flowers in panicles, appearing before the leaves; dioecious and without petals in ours; the fruit a terminally winged samara.

Key to the Species of *Fraxinus* in British Columbia
1a Twigs and leaves glabrous except for hairs along the undersides of the midveins of the leaflets. Leaflets distantly serrate. Buds black. Calyx none . . . . *F. excelsior*
1b Twigs and leaves puberulent, at least when young; leaflets later becoming glabrous above but persistently hairy beneath. Leaflets entire or finely crenate-serrulate. Buds brown. Calyx present but minute, funnel-shaped . . . . *F. latifolia*

Fraxinus latifolia Bentham   Oregon Ash
(*F. oregona* Nuttall)

Tree up to 20 m tall; twigs grey, puberulent, becoming glabrous; terminal bud conspicuous, tapering, brown, hairy. **Leaves** 20-30 cm long, with 5-7(-9) leaflets that are ovate to obovate, 5-15 cm long, acute to acuminate at apex and entire or minutely crenate-serrulate; sessile or nearly so, except for the terminal leaflet; puberulent, becoming glabrescent above but persistently hairy beneath. **Flowers** in dense axillary panicles, dioecious, with minute, funnel-shaped calyces but no petals. Staminate flowers have 2 stamens with anthers longer than the filaments, and no pistil, pistillate flowers have a minute ovary with a longer, forked style and stigma. **Fruit** samaras 3-5 cm long, the wing oblanceolate to spatulate, 3-9 mm wide, extending halfway down the sides of the ovary. Plate 68, figure 1. Photo 82.

Range: Moist (and sometimes dry) sites along the Pacific coast, from Seattle southward to California. There have been many reports of this as a native species on southern Vancouver Island, but few of them have been backed up by convincing specimens or firm information on their status. There are a number of *F. latifolia* trees in the city of Victoria, including Beacon Hill Park, but all those I have seen are standing in artificial surroundings (lawns, street verges, etc.), leaving me to wonder about their origins. There is also a recent record of one growing in natural surroundings at Macktush Creek, south of Port Alberni.

Fraxinus excelsior L.   **European Ash**

Grey-barked tree up to 30 m (or more) tall; twigs glabrous, grey; terminal buds stoutly pyramidal, black or almost so, glandular-roughened. **Leaves** 18-27 cm long, with 9-11 leaflets that are sessile except for the terminal one, ovate to lanceolate, acuminate, serrulate and glabrous except for hairs along the midvein beneath. **Flowers** staminate or perfect, with neither calyx nor corolla. **Fruit** with wings extending almost to the base and often with a twist. Plate 68, figure 2.

Range: Native of Europe. Commonly planted here, especially on southern Vancouver Island, and escaping locally, as in the Victoria area. This is the Ash that is now most likely to be encountered here.

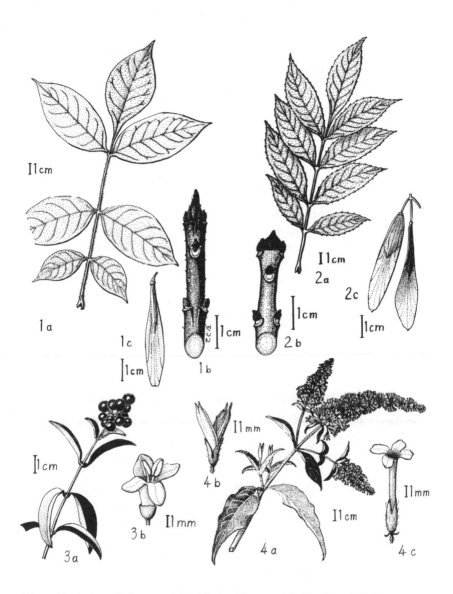

Plate 68 Ashes, Privet and Buddleja. Figures: (1) *Fraxinus latifolia;*
(2) *Fraxinus excelsior;* (3) *Ligustrum vulgare;* (4) *Buddleja davidii.*
1a, 2a, leaves; 1b, 2b, twigs and buds; 1c, 2c, 4b, fruit; 3a, fruiting branch;
4a, flowering branch; 3b, 4c, flowers.

Loganiaceae   Logania Family

Shrubs and trees with mostly opposite simple stipulate leaves. Flowers perfect and radially symmetrical, in panicles or cymes. Calyx and corolla 4- or 5-lobed. Stamens 4 or 5. Ovary superior, 2-chambered, with many ovules.

Buddleja   Butterfly-bush, Buddleja

Shrubs with stellate, glandular or scaly pubescence; winter buds acute and superposed (2 or more, one above another, in a leaf axil), with 2 outer scales. Leaves usually opposite. Flowers in panicles or compound spikes. Calyx and corolla 4-lobed. Fruit a 2-valved capsule with many minute seeds.

Buddleja davidii Franchet   Buddleja
Orange-eye Butterfly-bush

Shrub up to 5 m tall, with ascending and arching branches. Branches 4-angled and pubescent with stellate hairs. **Leaves** opposite, ovate to lanceolate, (3-)10-25 cm long, cuneate-based, serrate; dark green and soon glabrous above, white-felted beneath. **Flowers** in conspicuous terminal erect or arching, elongate panicles; calyx pubescent; corolla pale pink to deep purple, with orange in the throat, the tube about 1 cm long, the spreading lobes shorter; stamens 4, with very short filaments, inserted in the corolla tube between one-half and two-thirds of the distance between the base and the throat; style included. Ovary glabrous. **Fruit** an acute 2-valved capsule, 5-8 mm long, containing many seeds. Plate 68, figure 4.

Range: Native of eastern Asia; commonly cultivated here, and escaping to open stony sites, including gravel bars, around the Strait of Georgia, including the Gulf Islands, and in the lower Fraser River valley. There is appreciable variation in surface texture and other details among the various local, naturalized populations, reflecting the diverse cultivated horticultural forms that were their progenitors.

Polemoniaceae   Phlox Family

Mostly herbaceous, with few shrubby species. Flowers with petals joined into a tube, from the top of which the petal blades spread wide. Stamens 5. Ovary superior, 3-carpelled, with a style and 3 stigmas. Fruit a 3-chambered capsule.

Key to the Genera of Polemoniaceae
1a   Leaves opposite, linear to lanceolate, undivided; the bases of each pair joined around the stem by a white sheath .... *Phlox*
1b   Leaves mostly alternate, palmately divided .... *Leptodactylon*

Phlox   Phlox

Opposite-leaved, mostly herbaceous plants with flowers in terminal cymes. Calyx tube with green ribs projecting into acuminate teeth and separated by colourless membranes. Five stamens attached at different levels in the corolla tube. The following species are woody-based with herbaceous branches.

Key to the Species of *Phlox*
1a   Plant erect and more or less open in structure, with well-developed internodes; leaves 2-8 cm long.
     2a   Leaves linear; colourless membranes of calyx keeled near base; style 6-15 mm long .... *P. longifolia*
     2b   Leaves lanceolate; colourless membranes of calyx flat or almost so; style 0.5-2 mm long .... *P. speciosa*
1b   Plant low, dense and compact; cushion-forming, with very short internodes; leaves 13 mm or less long .... *P. caespitosa*

Phlox caespitosa Nuttall   Tufted Phlox

Low, densely branched, cushionlike, greyish subshrub from a taproot. Stems pale, densely glandular-hairy. **Leaves** opposite, usually less than 13 mm long, needlelike, glandular, with tufts of smaller leaves in their axils. **Flowers** usually single, on short pedicels terminating branches; calyx

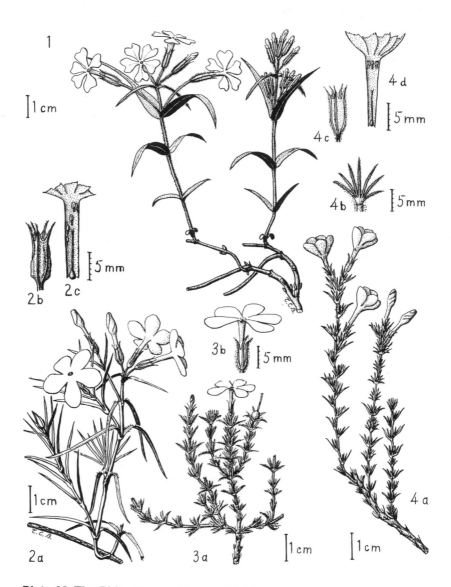

Plate 69 The Phlox Family. Figures: (1) *Phlox speciosa;* (2) *Phlox longifolia:* 2a, flowering branch; 2b, calyx; 2c, corolla tube opened out; (3) *Phlox caespitosa:* 3a, flowering and fruiting branch; 3b, flower; (4) *Leptodactylon pungens:* 4a, flowering branch; 4b, dorsal view of leaf; 4c, calyx; 4d, corolla tube opened out.

glandular-hairy, with flat, colourless intervals; corolla tube 8-14 mm long, less than twice as long as the calyx; corolla limb 15-20 mm wide when open; style 3-8 mm long. Plate 69, figure 3. Photo 83.

Range: In open grassy Ponderosa Pine parkland, in southern British Columbia, Washington, Idaho and Montana. In British Columbia, mainly in the East Kootenay region, northward to Columbia Lake; with one record from Walhachin, in the Thompson River valley.

Phlox longifolia Nuttall   Long-leaved Phlox

Low, ascending shrub up to 50 cm tall, with pale, almost shiny, glabrous twigs with well-developed internodes. **Leaves** linear, 1.5-8 cm long and 1-3 mm wide, acuminate-tipped, pale greyish green; glabrous except for a few cilia near the bases; the bases of a pair of leaves joined by a colourless membranous sheath around the stem. **Flowers** in loose terminal cymes, slender-pedicelled; calyx with the colourless intervals keeled toward the base; corolla tube 10-18 mm long; the limb 2-2.5 cm wide, white to pale pink; style 6-15 mm long. Plate 69, figure 2. Photo 77.

Range: Southern British Columbia to California in semi-arid steppe and open Ponderosa Pine parkland. In British Columbia, in the Okanagan and Similkameen valleys.

Phlox speciosa Pursh   Showy Phlox

Subshrub similar in structure to *P. longifolia,* up to 50 cm tall; stem pale green and minutely puberulent and glandular. **Leaves** lanceolate to linear, up to 7 cm long by 1 cm wide. **Flowers** have a glandular calyx, its colourless areas flat or almost so; corolla normally pink, the petals notched at their apices; style short, 0.5-2 mm long. Plate 69, figure 1.

Range: British Columbia to Montana and California. Rare in British Columbia, in dry grasslands and Ponderosa Pine parklands of the south Okanagan Valley.

Leptodactylon

Small shrubs with mostly alternate, palmately divided leaves with stiff, pungent (sharply pointed) divisions. Flowers on short, axillary, solitary branchlets, or apparently clustered at the ends of the branches by a shortening of the stem internodes; calyx similar to that in *Phlox;* 5 stamens with short filaments attached at the same level in the upper part of the corolla tube.

Leptodactylon pungens (Torrey) Nuttall

Small, spindly, aromatic, evergreen shrub, 10-60 cm tall, with stiff, brittle, sparingly branched stems; prickly by reason of the pungent leaf lobes. **Leaves** alternate or opposite (sometimes both on one plant), up to 1 cm long, palmately divided into 5-9 needlelike lobes. **Flowers** have calyces similar to that of *Phlox,* but with 5 pungent teeth; corolla 1.5-2.5 cm long, white or sometimes pale mauve or yellowish; opening wide at night, half-closed by day; the corolla tube and free petal blades each 8-12 mm long; stamens in a ring just within the throat of the corolla tube. Plate 69, figure 4.

Range: From the Okanagan Valley in British Columbia southward to Mexico and eastward to Montana, Nebraska and New Mexico; found in dry, sandy soils in semi-arid steppe.

Labiatae (Lamiaceae)   Mint Family

Aromatic herbs and shrubs, with stems mostly square in cross section. Leaves opposite, simple, the margins toothed. Flowers in terminal or axillary clusters; the corolla bilaterally symmetrical and 2-lipped; stamens 2 or 4; ovary superior, with 4 uni-ovuled chambers and a central undivided style. Fruit has 4 small nutlets.

Satureja   Savory

Flowers axillary. Calyx 12- to 15-veined. Corolla 2-lipped: the upper lip notched at tip and projecting forward; the lower lip 3-lobed and turned downward. Stamens 4, the lower pair longer than the upper.

Satureja douglasii (Bentham) Briquet   Yerba Buena

Prostrate subshrub with long, trailing stems. **Leaves** opposite, ovate, short-petioled, few-toothed, evergreen and sparsely hairy, the lower surfaces with scattered sessile glands. **Flowers** single (or rarely 2) on short pedicels in the leaf axils; corolla 7-10 mm long, white to purplish-tinged. Plate 71, figure 3.

Range: From southern British Columbia to California west of the Cascade Range; also in the southern Columbia Forest Region in the interior of British Columbia, from near Enderby, southward into Idaho; on rocky soils in open coniferous woods.

Solanaceae
Nightshade or Potato Family

Herbaceous or woody plants. Flowers with petals joined into a usually flaring tube, with the 5 corolla lobes spreading from the top. Stamens 5. Ovary superior, with 2 carpels and a headlike stigma. Fruit a capsule or berry.

Key to the Genera of *Solanaceae* in British Columbia
1a Shrubs with erect and arching, pale, spiny branches and axillary flowers .... *Lycium*
1b Straggling semi-climbers with green to greyish unarmed branches and cymes opposite the leaf axils .... *Solanum*

Lycium Matrimony-vine, Box-thorn

Spiny shrubs or trees with small, simple leaves and axillary clusters of 1-4 flowers. Stamens separate, with filaments longer than anthers. Calyx ruptured by the expanding fruit, which is a berry.

Lycium halimifolium Miller Box-thorn, China-berry

Shrub with long, arching, sparsely thorny branches with pale bark. **Leaves** alternate, simple, elliptic, entire, tapering to tips and to short petioles; pale green, with axillary tufts of small leaves and flowers, which may be on short spurs. **Flowers** 1-3 per cluster, on pedicels 1-2 cm long; corolla mauve with flaring tube 9-14 mm long and 5 spreading lobes about as long. **Fruit** an ellipsoid orange-to-red many-seeded berry, 1-2 cm long. Plate 70, figure 1.

Range: Native of Asia; cultivated and escaping in open semi-arid sites (Kamloops, Ashcroft, Keremeos); locally called China-berry.

Solanum   **Nightshade**

Herbs, shrubs or vines. Leaves alternate. Inflorescences appearing opposite the leaves, or terminal. Flowers with widely spreading corolla lobes, stamens with very short filaments, and longer anthers forming a conical cluster and opening by terminal pores. Fruit a berry.

Solanum dulcamara **L.**   **Bittersweet**

Extensively scrambling, more or less woody plant with pliable stems and green or sometimes purplish branches. Stems and leaves glabrous to thinly hairy. **Leaves** ovate, acuminate, entire or basally pinnately lobed, dark green above, sometimes tinged with blue. **Flowers** in cymes that are terminal or in positions other than in leaf axils; corolla lobes 3-4 times as long as the flaring tube and purple or occasionally white; anthers longer than the filaments and cohering into a bright yellow cone. **Fruit** a bright red ellipsoid berry. Plate 70, figure 2.

Range: Native of Europe. Naturalized in ditches and roadsides in widely scattered places in southern British Columbia (Hope, Trail, Vancouver, Victoria).

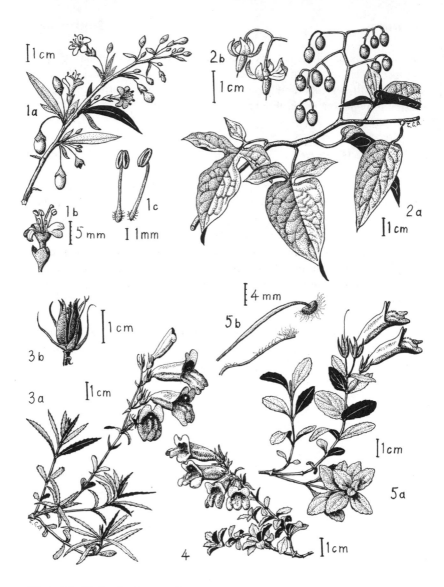

Plate 70 Nightshade and Figwort Families. Figures: (1) *Lycium halimifolium:* 1a, flowering and fruiting branch; 1b, flower; 1c, stamens; (2) *Solanum dulcamara:* 2a, fruiting branch; 2b, flowers; (3) *Penstemon fruticosus:* 3a, flowering branch; 3b, capsule; (4) *Penstemon davidsonii;* (5) *Penstemon ellipticus:* 5a, flowering branch; 5b, fertile and sterile stamens.

316 — SOLANACEAE AND SCROPHULARIACEAE

Scrophulariaceae   Figwort Family

Herbs, shrubs or trees. Leaves opposite or alternate. Flowers perfect; the corolla usually has 5 united petals, and is bilaterally symmetrical and bell-shaped to tubular, its limb 4- to 5-lobed, or 2-lipped with the lips 2- or 3-lobed; stamens usually 4, in unequal pairs, sometimes 2 or 5; ovary superior. Fruit a 2-chambered capsule or a berry (60).

Penstemon   Penstemon

Mostly herbaceous, but 3 of our species are low, sprawling evergreen or semi-evergreen shrubs with leathery, opposite leaves. Flowers lavender to purple, in terminal racemes in our shrubby species; calyx 5-lobed; corolla elongate, 2-lipped, the upper lip 2-lobed, the lower 3-lobed; stamens 5, one sterile and lacking an anther; anthers woolly in our shrubby species; the 2 anther sacs joined apically across the filament tip. Fruit a 2-chambered capsule with numerous seeds. A variable genus, typically of open rocky sites.

Key to the Shrubby Species of *Penstemon*
1a Leaves elliptic to obovate, 5-25 mm long and 1-2.5 times as long as wide. Stems prostrate to decumbent, forming mats with scattered erect flowering shoots; usually less than 15 cm high.
    2a Leaves on flowering stems small, often bractlike, rarely 1 cm long
    . . . . *P. davidsonii*
    2b Leaves on flowering stems well developed, usually at least 1 cm
    long . . . . *P. ellipticus*
1b Leaves lanceolate, 20-50 mm long and 3-10 times as long as wide. Stems decumbent to ascending, often more than 15 cm high
    . . . . *P. fruticosus*

Penstemon davidsonii Greene   Davidson's Penstemon
(*P. menziesii* Hooker: *nomen illegitimum*)

Prostrate shrub forming dense mats with flowering stems usually less than 15 cm high. **Leaves** elliptic to obovate, obtuse at tip, thick, firm, glabrous, 5-10(-12) mm long and at least half as wide. **Flowers** 1-4 in short racemes; calyx 7-10 mm long, slightly hairy to glabrous, the lobes lanceolate; corolla lavender to purple, 20-35 mm long; sterile stamen slender, usually bearded, shorter than the fertile stamens. Plate 70, figure 4. Photo 92.

Range: From British Columbia to California through the Coast, Cascade and Sierra Nevada ranges; mostly alpine, on rocks and ledges. In British Columbia, on Vancouver Island and in the Cascade and Coast ranges northward at least to the Skeena River valley. Most of our material is var. *menziesii* (Hooker) Cronquist, with serrulate leaf margins, which is predominantly on Vancouver Island. Var. *davidsonii*, with entire leaf margins, is common in the Cascade and Coast ranges northward to Mount Waddington and occasionally on Vancouver Island.

Penstemon ellipticus Coulter & Fisher   Oval-leaved Penstemon

Similar to *P. davidsonii,* but coarser in structure. **Leaves** elliptic, rounded to obtuse at apex, 10-15 mm long, distantly serrulate to nearly entire. **Flowers** in a strongly glandular-hairy raceme; calyx 8-15 mm long; corolla deep lavender, 27-40 mm long. Plate 70, figure 5.

Range: Southeastern British Columbia and southwestern Alberta, and in Montana and Idaho. In British Columbia, in rocky sites at subalpine and alpine elevations, in the Selkirk and Rocky Mountains, north to Yoho, Glacier and Mount Revelstoke national parks.

Penstemon fruticosus (Pursh) Greene   Shrubby Penstemon
(*P. scouleri* Lindley)

Sprawling shrub with stems decumbent to ascending, commonly 15-40 cm high. **Leaves** lanceolate, 20-50 mm long and 3-10 as long as wide, serrate to entire, acute at apex. **Flowers** 5-11 in an often 1-sided raceme; calyx 7-15 mm long, the lobes lanceolate, glandular-hairy; corolla lavender to purple, 30-50 mm long. Plate 70, figure 3. Photo 78.

Range: From southern British Columbia and southwestern Alberta southward to Oregon and Wyoming. In British Columbia, in open to sparsely wooded areas at moderate to alpine elevations in the southern interior; from Hope to the Crowsnest Pass and northwestward to Chezacut on the upper Chilcotin River.

Note: Most of our material belongs to var. *scouleri* (Lindley) Cronquist, with narrowly lanceolate leaves 6-10 times as long as wide, which ranges across the province. Var. *fruticosus,* with wider, often entire leaves, 2-7 times longer than wide, occurs in Manning Provincial Park.

Caprifoliaceae   Honeysuckle Family

Shrubs, vines, herbs or (rarely) trees, with opposite leaves. Flowers perfect; calyx 3- to 5-lobed or toothed; corolla of united petals, 4- to 5-lobed, radially or bilaterally symmetrical; stamens 5 or 4; ovary inferior. Fruit a drupe, berry or achene.

Key to the Genera of Caprifoliaceae

1a Leaves pinnately compound . . . . *Sambucus*
1b Leaves simple.
    2a Vines with entire-margined leaves . . . . *Lonicera*
    2b Shrubs.
        3a Stems erect.
            4a Leaves coarsely toothed and often 3-lobed. Flowers white in
            a flat panicle . . . . *Viburnum*
            4b Leaves entire or sometimes pinnately lobed. Flowers not
            white and not in flat panicles.
                5a Flowers in axillary pairs; corolla yellow, bilaterally sym-
                metrical. Fruit red or black . . . . *Lonicera*
                5b Flowers in compact clusters; corolla pink, usually radially
                symmetrical. Fruit white . . . . *Symphoricarpos*
        3b Stems trailing or arching.
            6a Vegetative stems quite prostrate. Leaves shallowly few-
            toothed. Flowers in terminal pairs on erect branchlets; corolla
            10-15 mm long. Fruit an achene . . . . *Linnaea*
            6b Vegetative and flowering stems often arching. Leaves entire
            or sometimes pinnately lobed. Flowers in terminal or axillary,
            dense clusters (very short racemes). Corolla 3-5 mm long. Fruit
            a white berry . . . . *Symphoricarpos*

Linnaea   **Twinflower**

Trailing subshrub with very slender, scarcely woody, weakly rooting, thinly hairy stems. Leaves opposite, simple, obovate to nearly circular, few-toothed, evergreen and leathery; the short petioles and leaf bases ciliate. Flowers in pairs on erect, forked peduncles terminating short, erect branchlets bearing 2-4 pairs of leaves; the flower base between 2 bractlets;

calyx 5-lobed; corolla slenderly funnel-shaped, 5-lobed at the mouth, pink (usually) to white, hairy in the throat; stamens 4; ovary inferior, with 3 carpels, of which only 1 is fertile; style about as long as the corolla, with a tiny, knoblike stigma. Fruit a 1-seeded achene, flanked by a pair of glandular bractlets.

Linnaea borealis L.   Twinflower

The only species. Plate 71, figure 4. Photo 84.

Range: Circumboreal, with several regional varieties (or subspecies, according to choice); in coniferous woods, often forming extensive loose mats on rotting logs.

Key to the Varieties of *L. borealis* in British Columbia
1a Leaves usually broadly obovate to nearly orbicular, obtuse at apex; calyx lobes short and blunt; corolla strongly flaring, short. Transcontinental, including the interior of British Columbia
. . . . **var.** *americana* **Rehder**
1b Leaves more narrowly elliptic to obovate, often acute at apex; calyx lobes attenuate; corolla longer and more slender. Coastal from southeastern Alaska to Oregon . . . . **var.** *longiflora* **Torrey**

Viburnum

Shrubs or small trees that have winter buds with 2 united outer scales. Leaves opposite and simple. Flowers in flattish cymes; calyx 5-lobed; corolla, in ours, white with spreading lobes; stamens 5; ovary inferior, unequally 3-chambered, only 1 chamber with an ovule; stigmas 1-3. Fruit a single-seeded drupe.

Key to the Species of *Viburnum*
1a Leaves shallowly 3-lobed and unlobed on the same branch. Inflorescence with all flowers alike . . . . *V. edule*
1b Leaves all prominently 3-lobed. Inflorescence with enlarged, sterile marginal flowers . . . . *V. trilobum*

Viburnum edule (Michaux) Rafinesque   Squashberry
(*V. pauciflorum* Pylaie)

Shrub 1-3 m tall, commonly sprawling, with glabrous branchlets. **Leaves** elliptic in outline, some of them shallowly 3-lobed above the middle; irregularly toothed, glabrous or slightly pubescent beneath, 5-8 cm long, often with a pair of glands at the junction of leaf blade and petiole. **Inflorescence** a cyme 1-3 cm wide, few-flowered. **Flowers** white, all alike and fertile; stamens shorter than corolla. **Fruit** red, 8-10 mm long with a flattened, scarcely grooved stone. Plate 71, figure 1. Photo 85.

Range: Transcontinental; in upland areas and along streams in woods throughout British Columbia.

Viburnum trilobum Marshall   High-bush Cranberry
(*V. opulus* L. var. *americanum* Aiton)
(*V. opulus* L. subsp. *trilobum* [Marshall] Hulten)

Shrub up to 4 m tall, with grey branches and glabrous branchlets. **Leaves** broadly obovate to nearly orbicular, all acuminately 3-lobed and coarsely dentate (though the middle lobe is sometimes elongate and entire), pilose on the veins beneath or nearly glabrous, 5-13 cm long; petiole with a shallow groove and small, usually stalked, glands. **Inflorescence** a cyme 8-10 cm wide, with sterile marginal flowers much larger than the fertile inner ones; stamens of fertile flowers about twice as long as corolla. **Fruit** scarlet and 8-10 mm long, with a flat smooth stone. Plate 71, figure 2. Photo 86.

Range: From the interior of British Columbia eastward to Newfoundland and southward to Washington state, Ohio and Pennsylvania. In British Columbia, rare and scattered in the southeastern interior from the southern Rocky Mountain Trench westward through the Columbia Forest Region to Westwold and the North Thompson River valley, northwestward to Houston. Found on riverbanks and in moist woods.

Plate 71 Figures: (1) *Viburnum edule:* 1a, flowering branch; 1b, twig and buds; 1c, flower; (2) *Viburnum trilobum:* 2a, flowering branch; 2b, fertile and sterile flowers (upper and lower views of the latter); (3) *Satureja douglasii:* 3a, flowering branch; 3b, flowers; (4) *Linnaea borealis.*

Lonicera Honeysuckle

Erect or climbing shrubs. Leaves opposite, entire, usually without stipules. Flowers in axillary pairs or terminal clusters. Calyx minute, 5-lobed. Corolla tubular or funnel-shaped, usually bilaterally symmetrical; 5-lobed, the lobes equal, or commonly 4 of them joined into an upper lip and the fifth forming a lower lip. Stamens 5, inserted on the tube of the corolla; ovary inferior, with 2 or 3 chambers; style slender; stigma knoblike. Fruit a berry with several seeds.

Key to the Species of *Lonicera*

1a Vines. Leaf pair below flowers united. Flowers in terminal clusters.
   2a Corolla tube 3-4 times as long as the short-lipped limb; stamens inserted below the tube mouth. Leaf margin ciliate .... *L. ciliosa*
   2b Corolla tube 1-2 times as long as the deeply lipped limb; stamens inserted at the mouth. Leaf margin normally not ciliate.
      3a Leaves deciduous, without stipules; bractlets a third as long as ovary .... *L. dioica*
      3b Leaves commonly evergreen, at least some pairs with connate stipules; bractlets half as long as ovary .... *L. hispidula*
1b Erect shrubs. Leaves all distinct. Flowers in axillary pairs.
   4a Flowers and black fruit with an involucre of conspicuous red or reddish bracts. Leaf tip acute to acuminate .... *L. involucrata*
   4b Flowers and red fruit with inconspicuous involucral bracts. Leaf tip obtuse to rounded .... *L. utahensis*

Lonicera ciliosa (Pursh) DC. Orange Honeysuckle

Vines with stems up to 6 m long, twining, trailing or climbing; stems slender, hollow and glabrous, and glaucous when young. **Leaves** 5-8 cm long, elliptic; rounded to pointed at apex, tapered to base; glabrous on both surfaces; green above, strongly bluish-glaucous beneath, the margins entire and ciliate; the uppermost leaves usually united around the stem to form a bowl-like involucre beneath the inflorescence. **Flowers** generally in a terminal compact spike or head or group of heads. Corolla orange to red, 2.5-4 cm long, the tube tapering, swollen at base on one side; the limb shortly 2-lipped, a quarter to a third as long as the tube; the upper lip 4-lobed, the lower lip narrow and 1-lobed. Stamens and style as long as the corolla; ovaries of the flowers distinct. **Fruit** a red berry. Plate 72, figure 1. Photo 91.

Range: British Columbia to California and Montana. In British Columbia, from the coast to the Rocky Mountains across the southern part of the province; northward to Comox, Quadra Island and Desolation Sound on the coast, and to Shuswap Lake in the interior; common in open woods and bush.

Lonicera dioica L.   **Red Honeysuckle**
var. *glaucescens* (Rydberg) Butters

Shrubs with trailing or climbing glabrous stems that are glaucous when young. **Leaves** 5-8 cm long, elliptic, almost sessile, glabrous and green above, whitened and thinly pubescent beneath; the margin normally not ciliate; the upper leaf pairs united around the stems. **Flowers** in small terminal heads. Corolla 2-3 cm long, yellow to purplish, becoming red; the tube 12-18 mm long, slender, slightly swollen at base on one side, often puberulent outside; lips nearly as long as the tube. Style longer than the corolla; it and the stamens exserted. **Fruit** red. Plate 72, figure 2.

Range: Eastern and northern British Columbia, eastward to Quebec and southward to Kansas. In British Columbia, in the Rocky Mountains and Rocky Mountain Trench; northward through the Peace River basin to the Northwest Territories boundary and westward up the Liard River to the Smith River.

Lonicera hispidula   **Purple Honeysuckle**
(Lindley) Douglas in Torrey & Gray   **Hairy Honeysuckle**

Stems slender, twining, 2-4 m long; young twigs usually short-hairy. Leaves ovate, often cordate at base, on petioles 3-5 mm long; green and glabrous or hairy above; glaucous and stiffly hairy, rarely glabrous, beneath; often evergreen; sometimes with broad joined stipules between the petioles; the upper leaf pairs united around the stem to form 1 or more nested involucres. **Flowers** in 1-5 terminal clusters; corolla 12-20 mm long, whitish to purple; 2-lipped, the lips as long as the tube, the lobes recoiled; the tube swollen at base on one side; stamens and style well exserted. **Fruit** red. Plate 72, figure 3.

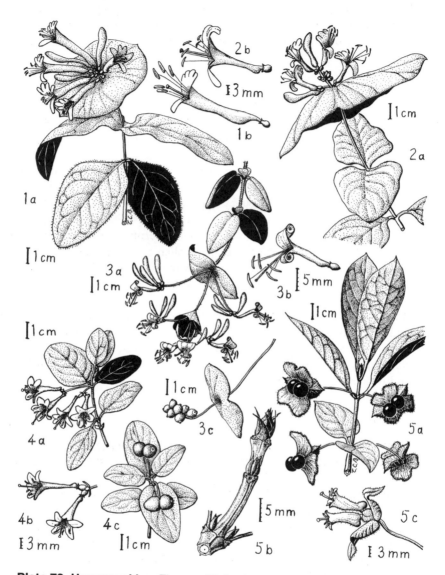

Plate 72 Honeysuckles. Figures: (1) *Lonicera ciliosa:* 1a, flowering branch; 1b, flower; (2) *L. dioica:* 2a, flowering branch; 2b, flower; (3) *L. hispidula:* 3a, flowering branch; 3b, flower; 3c, fruit; (4) *L. utahensis:* 4a, flowering branch; 4b, flowers; 4c, fruiting branch; (5) *L. involucrata:* 5a, fruiting branch; 5b, winter twig; 5c, flowers.

Range: British Columbia to California. In British Columbia, mainly on southern Vancouver Island and adjacent islands, and on the mainland coast north to Pender Harbour and Powell River; found on wooded rocky hillsides and in thickets.

Lonicera involucrata   Black Twinberry
(Richardson) Banks *ex* Sprengel

Shrubs 1-6 m tall, erect or sometimes leaning rather than twining their way up adjacent trees. Older stems up to 10 cm thick, with yellowish grey bark shredding into longitudinal strips. **Leaves** 5-14 cm long, elliptic to lanceolate, acute at apex, acute to rounded at base, entire, glabrous or nearly so, but with scattered sessile glands. **Flowers** in peduncled and involucrate pairs from the leaf axils; corolla yellow, cylindrical, nearly equally short-lobed; tube hairy, glandular and swollen on one side at the base; the floral bases concealed usually by the 4- to 6-bracted, hairy and glandular involucre, which is usually green at flowering time; stamens as long as corolla; style slightly exserted. **Fruit** paired, shiny black berries, loosely surrounded while young by the involucre, which darkens and reddens as it recurves from the ripe fruits. Plate 72, figure 5. Photo 87.

Range: Alaska to Quebec and southward to northern Mexico. Throughout British Columbia; common in moist wooded sites from sea level up to at least 1,500 m altitude in the south.

Lonicera utahensis Watson   Red Twinberry

Erect or ascending shrubs, 60-150 cm high; twigs glabrous or almost so. **Leaves** short-petioled, 2-8 cm long, thin, ovate, rounded at apex and base; usually glabrous above, thinly hairy beneath and on petiole; margins entire, ciliate near base and on petiole. **Flowers** paired, sessile on the ends of axillary peduncles, with or without a pair of minute linear bracts at the bases of the basally coherent ovaries; corolla 12-18 mm long, slenderly funnel-shaped, equal-lobed, swollen at base on one side, pale yellow; stamens about as long as corolla lobes; style exserted. **Fruit** a pair of red berries, joined to varying degrees, each about 7 mm across and containing 2-4 seeds. Plate 72, figure 4.

Range: Southern British Columbia to California and eastward to Alberta, Wyoming and Utah. In British Columbia, from the Coast Range near Cheekye to the Rocky Mountains; northward in the interior at least to Clearwater Lake and Anahim Lake (lat. 52°30'N). Typically found in medium to moist sites in the southern interior subalpine forest.

Symphoricarpos   Snowberries, Waxberries

Shrubs with very fine wiry twigs and simple, entire (or sometimes lobed), opposite leaves. Flowers small, in compact terminal and axillary clusters. Calyx 4- to 5-lobed. Corolla bell-shaped to funnel-shaped, commonly pale pink. Stamens 4-5, inserted on the corolla tube. Fruit a white 2-seeded berrylike drupe; subglobose, in tight clusters.

Key to the Species of *Symphoricarpos*
1a Corolla bell-shaped and hairy within.
   2a Style included (shorter than corolla), 2-3 mm long, glabrous.
      3a Trailing shrubs .... *S. mollis*
      3b Erect shrubs .... *S. albus*
   2b Style exserted (longer than corolla), 4-8 mm long, hairy
      .... *S. occidentalis*
1b Corolla narrower, funnel-shaped, often glabrous; style included, 2-4 mm long, glabrous .... *S. oreophilus*

Symphoricarpos albus (L.) Blake   Snowberry, Waxberry
(*S. racemosa* Michaux)

Densely colonial, profusely branched, erect shrub, 0.5-2 m tall; the older stems dark greyish brown with shredding bark; twigs thin, wiry, glabrous, yellowish brown. **Leaves** 2-5 cm long, ovate to elliptic, obtuse at base and apex, glabrous above, hairy beneath; entire or, on vigorous shoots, often sinuately pinnately lobed. **Flowers** short-pedicelled, in dense, short racemes terminating the branches and often also in the subterminal leaf axils. Corolla pink, widely bell-shaped, the tube often distended on one side, 5-7 mm long, hairy within; the lobes from half to fully as long as the tube. Stamens shorter than the corolla lobes; style 2-3 mm long, shorter than the corolla. **Fruit** densely clustered, white, 5-15 mm in diameter, variable in size even in one cluster; persisting on the bush through much of the winter. Plate 73, figure 1.

Key to the Varieties of *S. albus*
1a   Shrub 0.5-1 m high. Fruit 5-10 mm across . . . . **var.** *albus*
1b   Shrub 1 m or more high. Fruit 10 mm or more across
. . . . **var.** *laevigatus*

Range: A transcontinental species: var. *albus* from the interior of British Columbia to Quebec and southward to Virginia; var. *laevigatus* (Fernald) Blake from southern Alaska to Alberta and southward to California. The species is found almost throughout British Columbia, northward at least to the Sheslay River, north of Telegraph Creek, in moderately dry, open woods at low to moderate altitudes. Most of our material belongs to var. *laevigatus.*

Symphoricarpos mollis **Nuttall**	**Creeping Snowberry**
var. *hesperius* **(G.N. Jones) Cronquist**	**Trailing Snowberry**
in C.L. Hitchcock et al.	

A trailing shrub, its reclining stems rooting at the nodes, forming small tubers and giving rise to erect branches up to 60 cm high; twigs slightly pubescent. **Leaves** short-petioled, the blade ovate to elliptic, 1-3 cm long, slightly pubescent at least beneath, entire or sometimes lobed. **Flowers** in short terminal racemes, with sometimes a few flowers in the upper leaf axils. Corolla 3-5 mm long, widely and often asymmetrically bell-shaped, with lobes almost equal in length to the tube and with a collar of short hairs around the top of the corolla tube inside. Stamens as long as the corolla lobes; style 2-3 mm long, about as long as the corolla tube. **Fruit** 5-6 mm long. Plate 73, figure 3. Photo 88.

Range: Southern British Columbia to California and Idaho. In British Columbia, on southern Vancouver Island, northward to Port Alberni, Campbell River and Strathcona Park, and on the Gulf Islands. Found at low to middle elevations in dry semi-open rocky sites.

Symphoricarpos occidentalis Hooker

Wolfberry
Western Snowberry

A low, extensively colonial shrub, up to a metre high, spreading by rhizomes or rooting branches. Stems dark grey, with shredding bark; twigs reddish brown, puberulent. **Leaves** similar to those of *S. albus*, but apex sometimes mucronate; texture thicker and more leathery, puberulent at least beneath. **Flowers** sessile or nearly so, in dense short terminal or axillary spikes; corolla 5-8 mm long, bell-shaped, the prominent lobes somewhat flared; stamens and style equalling the corolla or slightly longer and exserted; the style hairy at the middle or, in ours, often glabrous. **Fruit** 6-9 mm long. Plate 73, figure 2.

Range: Widespread from the interior of British Columbia across the prairie provinces to Manitoba, and introduced in Ontario, and southward through the central United States to Kansas and New Mexico. In British Columbia, in the Peace River basin, and in the East Kootenay region north to Windermere and westward and northward to the Okanagan Valley, Chilcotin Plateau, Vanderhoof, the Stikine Valley and the Toad River. It is found in dry grassland and deciduous woods, often in hollows where snow can accumulate. It tends to replace *S. albus* in semi-arid areas.

Symphoricarpos oreophilus Gray
var. *utahensis* (Rydberg) Nelson
in Coulter & Nelson

Mountain Snowberry

Erect shrub up to 1.5 m tall; twigs pubescent (in our material) or glabrous. **Leaves** on petioles up to 3 mm long; the blade ovate to elliptic, usually pubescent, sometimes glabrous. **Flowers** on short, drooping pedicels, in short terminal racemes or in 1- to 3-flowered axillary and terminal clusters; corolla narrowly funnel-shaped, 7-11 mm long, usually hairy within; the lobes a quarter to half as long as the tube; stamens shorter than the corolla lobes; style glabrous, shorter than the corolla tube. **Fruit** white, 7-10 mm long. Plate 73, figure 4.

Range: Southeastern British Columbia to Utah; var. *oreophilus* from Utah and Colorado southward to northern Mexico. In British Columbia, found so far only from the southern Monashee Mountains near Grand Forks and in the southern Rocky Mountain Trench near the United States border.

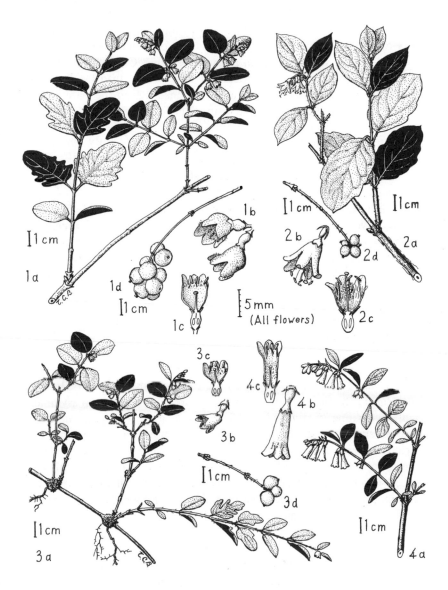

Plate 73 Snowberries. Figures: (1) *Symphoricarpos albus;*
(2) *S. occidentalis;* (3) *S. mollis;* (4) *S. oreophilus.* 1a, 2a, 3a, 4a, flowering
branches; 1b, 2b, 3b, 4b, flowers; 1c, 2c, 3c, 4c, longitudinal sections of
flowers; 1d, 2d, 3d, fruits.

Sambucus   **Elderberries, Elders**

Shrubs or small trees, seldom herbs; stems with large pith; buds with 3-5 pairs of scales. Leaves opposite, pinnately compound; leaflets finely toothed. Flowers small, white, in terminal panicles. Calyx lobes minute; corolla 5-lobed, radially symmetrical, the lobes longer than the radius of their flattened combined base; stamens 5, inserted on the base of the corolla; stigmas 3. Fruit a berrylike drupe with 3 small stones.

Key to the Species of *Sambucus*
1a Inflorescence flat; fruit very glaucous . . . . *S. cerulea*
1b Inflorescence conical, ovoid or hemispheric; fruit lustrous, not glaucous.
  2a Inflorescence ovoid to conical, normally longer than wide. Fruit red to yellowish . . . . *S. racemosa*
  2b Inflorescence broadly hemispheric, wider than high. Fruit shiny black . . . . *S. melanocarpa*

Sambucus cerulea **Rafinesque**   **Blue Elderberry**
(*S. glauca* Nuttall)   **Blue-berry Elder**

Shrubs or small trees 3-10 m tall and up to 25 cm in trunk diameter, with finely fissured bark and glaucous twigs. **Leaves** with 5-9 leaflets; leaflets lanceolate to elliptic, strongly acuminate, sharply serrulate, often asymmetrical, occasionally divided, 7-15 cm long, glabrous, pale beneath. **Inflorescence** a broad flattened panicle, with 4 or 5 strong lateral branches around a shorter, slender central axis. **Flowers** white or creamy, 4-6 mm across; corolla lobes spreading; flowering from June to August. **Fruit** globose, 4-6 mm across, black with a dense bloom, giving it a pale powdery-blue appearance, edible; the stones wrinkled, with rough, irregular ridges and smooth intervals. Plate 74, figure 3. Photo 90.

Range: Southern British Columbia to Montana, California and New Mexico. On open well-drained ground or stream banks across southern British Columbia; on Vancouver Island northward to Campbell River and up to 600 m altitude; in the interior, northward to Savona and at altitudes up to 1,100 m.

Sambucus racemosa L.
subsp. *pubens* **(Michaux) House**
(*S. pubens* **Michaux)**

Red Elderberry
Red-berry Elder

Big coarse shrub up to 5 m tall, with soft, pithy, thick, commonly somewhat glaucous twigs. **Leaves** usually have 5-7 leaflets, which are elliptic to lanceolate, acuminate at apex, serrulate, 5-15 cm long and often asymmetrical at base. **Inflorescence** an ovoid to conical panicle, usually longer than wide, with short lateral branches on a distinct, stronger central axis, 6-10 cm long in flower, larger in fruit. **Flowers** white or creamy, 3-6 mm across; corolla lobes spreading, becoming reflexed; flowering in April and May. **Fruit** usually scarlet, occasionally varying to brown, yellow or white; not glaucous; 5-6 mm across. Plate 74, figure 1. Photo 89.

Key to the Varieties of *S. racemosa* in British Columbia
1a Fruit stones with smooth to pebbled surfaces. Tall shrub or tree 2-10 m
tall . . . . **var.** *arborescens*
1b Fruit stones more or less corrugated. Shrubs up to 3 m tall
. . . . **var.** *pubens*

Range: A circumboreal species of moist, shady wooded habitats. The completely glabrous subsp. *racemosa* occurs in Eurasia. The North American subsp. *pubens* (Michaux) Koehne includes several varieties that are usually pubescent to some degree. Var. *pubens* (Michaux) House is transcontinental and all over British Columbia, primarily in the interior but also locally on the coast, including Vancouver Island and the Queen Charlotte Islands, where it intermingles and intergrades in character with the weakly distinguished var. *arborescens* (Torrey & Gray) Gray. Var. *arborescens* is primarily coastal, from Alaska to California, and locally inland. In British Columbia it is common west of the Coast Range at medium to low elevations, and also occurs in the Prince George to Pine Pass area.

Sambucus melanocarpa Gray
(*S. racemosa* **var.** *melanocarpa*
[Gray] McMinn)

Black Elderberry
Black-berry Elder

Shrub 1-4 m tall, or maybe a small tree. **Leaves** rather shiny above, generally pubescent beneath. **Inflorescence** hemispheric, rounded on top, broader than long, rather umbel-like, with its central axis not clearly distinguished from the equally well-developed lateral branches. Corolla lobes spreading to somewhat reflexed. Flowering from May to July. **Fruit** shiny black or purple-black, the stones with finely wrinkled or pebbly surfaces. Plate 74, figure 2.

Range: From central British Columbia and western Alberta southward to Arizona and California at moderate to subalpine elevations. In British Columbia, mainly in the southeast, but northward in the Rocky Mountains to Pine Pass and westward to Hope; apparently not west of the Fraser River in the interior. Found in moist woods.

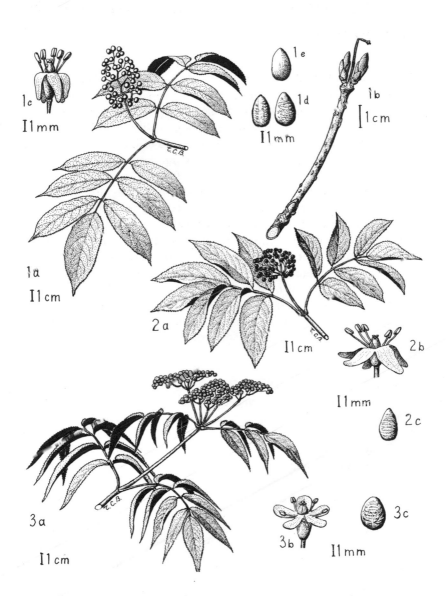

Plate 74 Elders. Figures: (1) *Sambucus racemosa;* (2) *S. melanocarpa;*
(3) *S. cerulea.* 1a, 2a, 3a, fruiting branches; 1b, winter twig;
1c, 2b, 3b, flowers; 1d, 1e, 2c, 3c, fruit stones (1d from *S. racemosa* var.
pubens, and 1e from *S. racemosa* var. *arborescens*).

ɔmpositae (Asteraceae)   Sunflower Family

Mostly herbs of the thistle, dandelion and sunflower type, but including a number of shrubs and trees. Leaves without stipules, variously arranged, simple but often deeply lobed or divided. The basic inflorescence unit is a head, with a number of small flowers borne sessile on a common flat or convex receptacle, surrounded by an involucre of bracts. Flowers of two main types in most species: a group of *disc flowers* occupying the surface of the usually disc-like receptacle, surrounded by a marginal series of *ray flowers*. Disc flowers have tubular, 5-lobed corollas; ray flowers are bilaterally symmetrical by a conspicuous one-sided extension of the corolla. Calyx in the form of a pappus (having bristles, scales or teeth) but sometimes absent. Stamens and pistils normally present, but either may be absent in ray flowers of a head in which the disc flowers are fertile. Stamens united at the anthers, which form a tube into which the pollen is shed, to be pushed out by the growing style. Ovary inferior; stigmas on the inner surfaces of 2 style branches, which are pressed together and appear as one in a young flower. Fruit an achene containing a single seed and usually crowned by the expanded pappus.

Key to the Shrubby Genera of Compositae

1a Flower heads many, in elongate racemes or compound racemes; pappus none; leaves oblanceolate, or broader and lobed or toothed
. . . . *Artemisia*

1b Flower heads in broad corymbose panicles (sometimes elongated in *Haplopappus*); pappus of fine bristles; leaves linear or narrowly lanceolate, entire.

   2a Involucre of 4-5 equal bracts . . . . *Tetradymia*

   2b Involucre of 10 or more overlapping bracts.

      3a Bracts overlapping in 5 vertical rows; ray flowers absent
. . . . *Chrysothamnus*

      3b Bracts not in vertical rows; ray flowers present . . *Haplopappus*

Artemisia

Aromatic shrubs or herbs. All our shrubby species have grey-hairy twigs and foliage. Leaves alternate, often dissected. Flowers all tubular disc flowers in small heads. All flowers perfect or the outer ones pistillate only. Corolla tubular, 5-toothed in the perfect flowers. Involucral bracts in several series, often with thin parchmentlike margins. Fruit a glabrous but often glandular achene without a pappus.

Key to the Shrubby Species of Artemisia

1a Leaves all entire .... A. cana
1b Leaves, at least on vegetative shoots, toothed or lobed.
    2a Leaves cuneate, 3-toothed across the truncate apex
       .... A. tridentata
  2b Leaves dissected, at least on vegetative shoots.
    3a Leaves of vegetative shoots normally 3-lobed near apex; receptacle glabrous .... A. tripartita
    3b All leaves pinnately divided and at least the upper lobes again divided; receptacle hairy .... A. frigida

Artemisia tridentata Nuttall   Big Sagebrush

A pale grey, much-branched shrub, 0.5-3 m tall, highly variable; with dark grey to blackish, shredding bark on old stems; the young twigs covered with closely matted white hairs. **Leaves** 1-4 cm long, usually cuneate, with 3 teeth across the truncate apex and narrowing to an almost sessile base; leaves in the inflorescence often entire, pale grey with appressed hairs on both surfaces. **Inflorescence** a long, erect terminal panicle of numerous heads. **Heads** 3-5 mm long, the outer bracts woolly, the inner glabrous; receptacle glabrous. **Flowers** 3-5 per head at low elevations, 5-8 at high altitudes; all perfect, with glandular yellow to greenish corollas; the 5-pointed tip of the anther tube shortly projecting from the corolla; style and stigmas exserted. **Fruit** a glandular achene. Plate 75, figure 1.

Range: British Columbia to Mexico and eastward to North Dakota. In British Columbia, common in the dry southwestern interior from the Okanagan Valley northward at least to the lower Chilcotin River, and in the Rocky Mountain Trench, northward to Windermere. Found typically on open dry grassy hills, reaching its best development on clay soils, as on

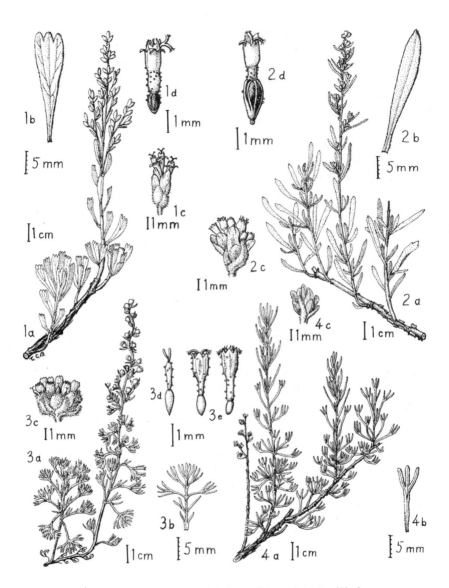

Plate 75 Sagebrushes. Figures: (1) *Artemisia tridentata;* (2) *A. cana;* (3) *A. frigida;* (4) *A. tripartita.* 1a, 2a, 3a, 4a, branches; 1b, 2b, 3b, 4b, leaves; 1c, 2c, 3c, 4c, heads of flowers; 1d, 2d, 3d, 3e, individual flowers (3d marginal pistillate flower, 3e perfect flowers).

ancient lacustrine terraces, where it may become a dominant plant under heavy grazing (Photo 93); ascending the mountains to 2,000 m altitude near Keremeos.

Key to the Varieties of *A. tridentata*
1a Involucres narrowly bell-shaped, about 4 mm high by 2 mm wide, with 3-5 flowers per head. Plants inhabiting arid areas at low altitudes
. . . . **var.** *tridentata*
1b Involucres wider: about 5 mm high by 4 mm wide, with 5-8 flowers per head. Plants inhabiting cooler, moister balds at higher altitudes (1,150-2,000 m) . . . . **var.** *vaseyana* **(Rydberg) Boivin**

Artemisia tripartita **Rydberg**   **Threetip Sagebrush**
(*A. trifida* **Nuttall not Turczaninow)**

Grey shrub 30-150 cm tall; stems dark grey with shredding bark; twigs often in broomlike bunches. **Leaves** very slender, and on vegetative branches deeply 3-parted into linear lobes, the lobes sometimes again divided. **Inflorescence** a slender panicle of heads. **Heads** 3-4 mm high, with woolly outer bracts and parchmentlike inner bracts, and a glabrous receptacle. **Flowers** 3-9 per head, yellow. Plate 75, figure 4.

Range: Southern British Columbia to California and Colorado in open dry grassland. Rare in British Columbia: found only in the Okanagan, Similkameen and Kettle valleys.

Artemisia cana **Pursh**   **Silver Sage, Hoary Sage**

A lower and more spreading shrub than *A. tridentata*, with shallow rhizomes. Branches pale grey, with loosely flaking bark; twigs whitish, densely short-hairy. **Leaves** narrowly oblanceolate, sessile, acute at both ends, usually entire, occasionally with 1-3 small teeth apically, white-pubescent all over. **Inflorescence** a slender leafy panicle of few to many heads. **Heads** 4-5 mm long, the outer bracts short and woolly; the inner bracts glabrous, parchmentlike and translucent; the receptacle glabrous. **Flowers** 6-20 per head, yellow, all perfect. Plate 75, figure 2.

Range: British Columbia to Saskatchewan and southward to California and New Mexico. Rare in British Columbia; reported from the southern

Rocky Mountain Trench (32) and from Logan Lake in the Nicola Basin. Typically a prairie species. The Logan Lake plants are on a roadside and may not be native there.

Artemisia frigida Willdenow   Pasture Sage

Low shrub, usually less than 40 cm tall; lower branches sprawling, slender flowering branches erect. **Leaves** soft, grey-hairy, finely pinnately dissected into narrow linear lobes, the larger lobes again divided. **Inflorescence** a very slender terminal panicle with erect branches, often longer than its supporting branch. **Heads** 3-4 mm high; 3-4 mm wide in the south, but up to 6 mm wide north of latitude 56°N (Peace River); bracts grey-hairy, the receptacle with hairs among the flowers. **Flowers** mostly perfect with yellow tubular glandular corollas and included anther tubes, but with a few marginal pistillate flowers with slender corollas, exserted style and stigmas, and no stamens. Plate 75, figure 3.

Range: Transcontinental: from Alaska to the Atlantic coast and southward through the Great Plains to Arizona; also in Siberia. In British Columbia, widespread in the drier areas of the interior, northwestward to Atlin and the Stikine River valley. A plant of dry grasslands, especially where disturbed; often invading overgrazed pastures and abandoned farmland.

Chrysothamnus   Rabbit-brush, Bigelowia

Shrubs with narrow alternate leaves and usually small heads of rayless yellow flowers in the autumn. Flower heads usually borne in broad terminal corymbose clusters; the involucre of many small lanceolate bracts arranged in 4 or 5 vertical series; receptacle naked; each head holding 5-20 flowers. Flowers all perfect, with yellow tubular corollas; the pappus a ring of fine, minutely hairy bristles. Fruit a narrow achene crowned with the pappus bristles. (*Bigelovia,* according to Henry [28], for *Bigelowia.*)

Key to the Species of *Chrysothamnus*
1a Stems pale, with matted white hairs . . . . *C. nauseosus*
1b Stems green, thinly tomentose or glabrous . . . . *C. viscidiflorus*

Chrysothamnus nauseosus **(Pallas) Britton**   **Rabbit-brush**
(*Bigelowia graveolens* **[Nuttall] Gray;**   **Golden-brush**
vide **Henry [28])**

Shrubs up to 120 cm tall, with light to dark grey lower stems and white-tomentose twigs. **Leaves** linear, 25-60 mm long, white tufted-tomentose. **Inflorescence** cymose or corymbose, broad and flattened or dome-shaped, compact. **Heads** numerous, rayless, 1 cm or more high; involucre narrowly bell-shaped, of 4-5 vertical series of narrow, acute, keeled, white-hairy bracts with thin, colourless margins, the outer ones white-tomentose. **Fruit** linear, appressed-hairy, crowned by a pappus of numerous soft bristles. Plate 76, figure 1. Photo 94.

Range: British Columbia to Saskatchewan and southward to northern Mexico. In British Columbia, in the dry southern interior, in grasslands and Ponderosa Pine parklands, northward to Kootenay National Park and to near Soda Creek on the Fraser River; a showy autumn flowerer, especially on stony or disturbed ground.

Chrysothamnus viscidiflorus   **Green Rabbit-brush**
(Hooker) Nuttall

Shrub up to 120 cm tall, densely branched and rather broomlike; twigs thinly tomentose to nearly glabrous, rather gummy, not whitened. **Leaves** linear-lanceolate, short-hairy and glandular, green, 2-4 cm long. **Heads** in a broadly domelike or flattened corymbose cluster with an involucre of glabrous bracts in more or less vertical rows. **Fruit** covered with appressed hairs and crowned with a pappus of fine bristles. Plate 76, figure 2.

Range: Southern British Columbia to North Dakota and New Mexico. Uncommon in British Columbia, near Osoyoos, on semi-open rocky slopes.

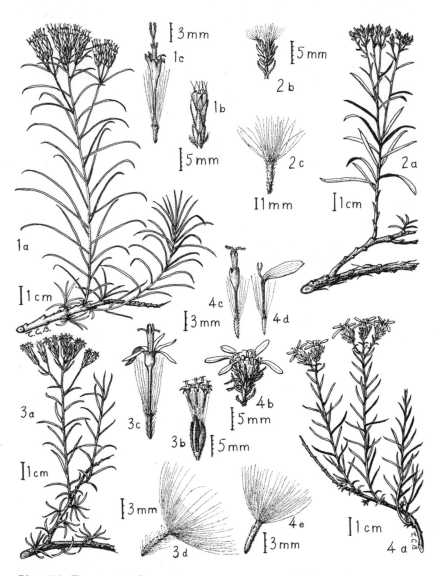

Plate 76 Figures: (1) *Chrysothamnus nauseosus;* (2) *Chrysothamnus viscidiflorus;* (3) *Tetradymia canescens;* (4) *Haplopappus bloomeri.*
1a, 2a, 3a, 4a, flowering branches; 1b, 2b, 3b, 4b, heads of flowers;
1c, 3c, 4c, tubular disk flowers; 4d, ray flower; 2c, 3d, 4e, achenes with pappus hairs.

Tetradymia   **Horsebrush**

Low rigid shrubs with stiff arching branches, sometimes spiny by modification of the leaves. Branches and leaves white-tomentose. Leaves alternate, entire and often bearing small tufts of secondary leaves in their axils. Heads with involucres of 4-6 concave, overlapping bracts, and containing about 4 tubular flowers and no ray flowers. Corolla elongate, with a slender cylindrical lower tube; the lanceolate lobes longer than the distinct short bell-shaped throat and equal in length to the projecting anther tube and style. Achenes short, round in cross-section, marked by 5 longitudinal lines and crowned by a fine white pappus.

Tetradymia canescens DC.   **Grey Horsebrush**

Pale shrub with spreading branches, 30-100 cm tall, spineless; branches densely and closely woolly. **Leaves** 1-3 cm long; linear (sometimes oblanceolate or spatulate) with an acute tip, narrowing to a sessile base, woolly, sometimes with fascicles of short, relatively wide leaves in the axils. **Inflorescence** a small, compact cymose cluster wider than high. **Heads** about 1 cm high; the involucre about two-thirds the length of the head; of 4 wide, bluntly keeled, thinly sericeous bracts. **Fruit** a tapering achene, short-hairy or glabrous, crowned by the spreading pappus. Plate 76, figure 3.

Range: Dry southern interior of British Columbia to Montana and New Mexico. Uncommon in British Columbia, confined to the Okanagan, Kettle, Thompson and Nicola valleys; usually on stony soils in the Ponderosa Pine – Bunch Grass zone of the Montane Forest – Grassland transition.

Haplopappus   **Bristleweed, Goldenweed**

Herbs or shrubs with yellow ray flowers as well as disc flowers. Heads with bracts of unequal length, the outer bracts intergrading with the reduced leaves on the peduncle, the overlapping bracts not arranged in vertical rows. Pappus of brownish bristles. Achenes covered with appressed hairs.

Haplopappus bloomeri Gray   Rabbit-brush, Goldenweed
(_Ericameria bloomeri_ [Gray] Macbride; _Chrysothamnus_
bloomeri [Gray] Greene)

Low shrub, 10-50 cm tall, with brown stem and twigs. Twigs and foliage
thinly clothed with very fine white cobwebby hairs and somewhat
gummy. **Leaves** sessile, linear, 1-6 cm long, tapering to apex and base,
green. **Inflorescence** a dome-shaped (or sometimes elongate) group of
heads terminating the stem and short lateral branches. **Heads** 7-10 mm
high, with around 20 overlapping hairy bracts, the inner bracts ciliate, and
a naked receptacle; and containing 10-20 flowers. **Flowers** of 2 kinds: the
majority are tubular perfect disc flowers with corollas around 8 mm long,
the anther tube protruding and styles with only the stigmatic branches protruding from the anther tube and bearing short hairs on their outer surfaces; the others, 1-5 marginal ray flowers in the head, have no anther
tube and the style protrudes up to 2 mm and bears glabrous stigmatic
branches. **Fruit** an achene 4-5 mm long with appressed white hairs; pappus bristles brownish and of unequal lengths, around 5-7 mm. Plate 76,
figure 4.

Range: Southern British Columbia to California. Rare in British Columbia, near Westbridge, in the Kettle River valley, and Keremeos. Found in
open woods on dry rocky slopes.

GLOSSARY

Acuminate Having a long tapering apex (Plate 79, Leaf Apex).

Acute Ending in an apex of less than 90° (Plate 79, Leaf Apex).

Achene A small dry one-seeded indehiscent carpel or fruit.

Alternate With a single leaf or bud at each node along a stem (see Opposite).

Anther The pollen sac on the stamen (Plate 78, figures 3 and 4).

Apical Of or at the apex.

Apiculate Terminated by a short acute tip.

Appressed Lying close and flat against.

Arcuate (veins) Lateral veins curving to become parallel to leaf margin.

Areole In cactuses, equivalent to the node of a conventional stem; the point where spines emerge.

Awn A bristlelike appendage.

Axil The upper angle between a leaf and the stem (Plate 78, figure 1).

Biserrate Coarsely serrate, bearing small teeth along the margins of the larger teeth (Plate 79, Leaf Margin).

Bloom Grey, powdery surface material, which can be easily rubbed off; e.g., the delicate powder on plums.

Bract A modified leaf with a flower or inflorescence arising from its axil (Plate 78, figure 3). In conifer seed-cones, a modified leaf with a seed-bearing cone-scale arising from its axil (Plate 77, figure 1).

Bractlet A small or secondary bract.

Bundle (1) In some conifers (e.g., pines and larches), a group of leaves. (2) In plant anatomy, the water- and food-conducting strands in the roots, stems and leaves, including the veins in leaves.

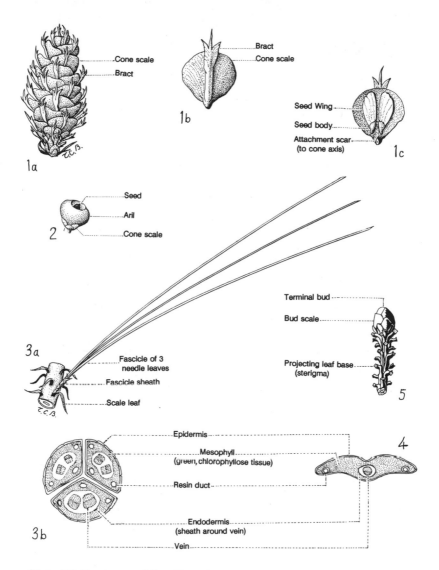

Plate 77 Anatomy of Conifers: Leaves, Twigs, and Cones. Figures:
(1) *Pseudotsuga menziesii:* 1a, seed cone; 1b, cone scale, dorsal view,
showing bract; 1c, cone scale, ventral view, with seeds; (2) *Taxus
brevifolia,* seed-bearing cone; (3) *Pinus ponderosa:* 3a, leaf fascicle;
3b, cross section of leaf fascicle near base; (4) *Abies amabilis,* cross
section of leaf; (5) *Picea sitchensis:* twig and bud.

Bundle scar The mark on a leaf scar where a bundle or vein to the leaf broke off when the leaf fell (Plate 78, figure 2).

Calcareous Containing a high proportion of calcium carbonate (lime); chalky.

Calyx (plural: calyces) The outer floral envelope, composed of sepals.

Capsule A dry fruit, of more than one carpel, that opens upon maturity.

Carpel A modified leaf, rolled up to form a chamber (ovary) for ovules and seeds; the tip forms the style and stigma. One or more joined carpels make up a pistil.

Catkin A scaly-bracted spike of minute unisexual flowers, commonly hanging and pollinated by wind (Plate 79, Inflorescences).

Ciliate Fringed with marginal hairs.

Coetaneous Flowering at the same time as leaf expansion.

Colonial Forming spreading colonies by sprouting from rhizomes or roots.

Compound Composed of two or more similar parts; e.g., the leaflets of a compound leaf.

Connate Joined together, as of similar organs.

Cordate Heart-shaped, usually with the stem or attachment at a sinus between the two lobes (Plate 79, Leaf Base).

Corolla The inner floral envelope, composed of petals.

Corymb An inflorescence, commonly flat-topped or convex, with pedicels of different lengths, whose terminal flower opens last (Plate 79, Inflorescences).

Corymbose Like a corymb.

Cotyledon Leaves developed in the embryo, often appearing above ground after germination.

Crenate With rounded teeth (Plate 79, Leaf Margin).

Cuneate Wedge-shaped (Plate 79, Leaf Base).

Cyme A flat-topped or convex flower cluster whose central or terminal flower opens first.

Cymose Like a cyme.

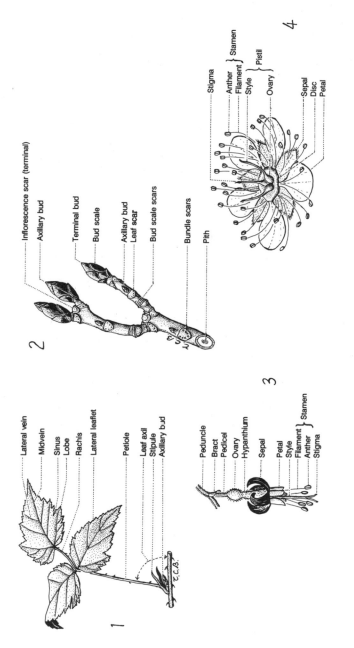

Plate 78 Anatomy of Flowering Trees and Shrubs: Leaves, Twigs, and Flowers. Figures: (1) compound leaf of *Rubus ursinus*; (2) winter twig of *Aesculus hippocastanum*, with buds; (3) flower of *Ribes lobbii*, with inferior ovary; (4) flower of *Physocarpus capitatus*, with superior ovaries.

Deciduous Falling in the year produced.

Decumbent Lying along the ground with the end ascending.

Dehiscent (of seed pods) Opens to release seeds.

Dentate Toothed, with the teeth directed outward.

Depressed-globose An oblate spheroid; shaped like a globe that is flattened at the base and apex, like a mandarin orange.

Dioecious Having staminate and pistillate flowers on different plants of the same species.

Disc A nectar-secreting pad surrounding the base of the ovary in a flower (Plate 78, figure 4).

Distantly Referring to the teeth on a leaf margin, spaced far apart relative to their size.

Drupe A cherrylike fruit that contains a stone.

Drupelet One of an aggregate of small drupes.

Entire Having a smooth, toothless margin (Plate 79, Leaf Margin).

Erose Having a jagged margin, as though worn thin.

Evergreen Leaves that remain green and persist on the twig for more than one year. A plant that retains green leaves throughout the year.

Exfoliating Peeling off in thin layers.

Exserted Projecting beyond the enclosing floral appendages.

Fascicle Compact axillary cluster (Plate 77, figure 3).

Filament Any threadlike body; specifically, the part of the stamen that supports the anther (Plate 78, figures 3 and 4).

Foliolate Having leaflets: trifoliolate, with 3 leaflets; 5-foliolate, with 5 leaflets.

Follicle A dry dehiscent fruit opening along one seam.

Free Unattached.

Genus (pl. genera) A named group of related species; the name that precedes that of the species; e.g., *Acer,* the maples.

Glabrescent Becoming glabrous with age.

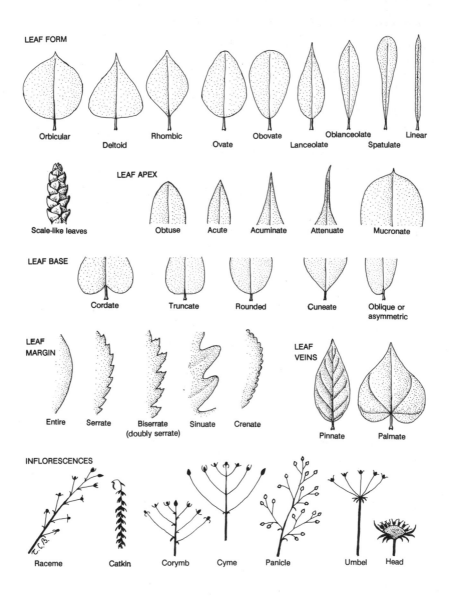

Plate 79 Leaf Forms and Inflorescences.

Glabrous Without hairs; smooth.

Glandular Bearing glands or hairs that secrete a viscid fluid.

Glaucous Covered with a bloom, whitened.

Globose Globe-shaped; spherical, but without the mathematical precision of a sphere. (See Depressed-globose.)

Habit The overall growth form of a plant.

Hip The fruit of a rose: several achenes enclosed in a flasklike calyx cup.

Hypanthium The part of the cuplike or tubelike flower base that does not adhere to the ovary and bears the free tips of the sepals, petals and stamens on its rim (Plate 78, figure 3).

Intergrading With overlapping ranges of variation; merging together by gradual blending of characters.

Included Within surrounding organs, not exserted.

Inferior (of the ovary) Below the flower parts, united to the calyx.

Inflorescence A flower cluster.

Infrastipular Below the stipules.

Internode The segment of stem between successive nodes.

Involucre A circle of distinct or united bracts subtending a flower or cluster of flowers.

Involucrate Having an involucre.

Krummholz A low, contorted, shrubby form of a species that is normally a tree. Krummholz forms are found in sites exposed to strong winds, usually at high altitudes.

Lanceolate Lancelike, three or four times as long as the widest part, which is at or below the middle (Plate 79, Leaf Form).

Leaflet One of the blades of a compound leaf, with no bud in its axil.

Leaf scar The scar left on a twig after a leaf has fallen (Plate 78, figure 2).

Lenticels Tiny raised pores on the stem, through which gaseous exchange occurs.

Limb The expanded outer part of a corolla of joined petals, commonly where the petal ends are separate (as in *Phlox*).

Locule The chamber in an ovary containing the ovules, and later the seeds.

Loculicidal Dehiscent through the seed chambers (of a capsule).

Monoecious Having staminate and pistillate elements in separate flowers on the same plant.

Mucronate Ending in an abrupt but distinct tip (Plate 79, Leaf Apex).

Node The point on a stem where one or more leaves are attached.

Oblanceolate Lancelike (lanceolate), but widest toward the apex and more sharply pointed at the base (Plate 79, Leaf Form).

Oblong-lanceolate Narrow and pointed, with nearly parallel sides .

Obovate Inversely egg-shaped in outline, widest near the apex (Plate 79, Leaf Form).

Opposite Arrangement of two leaves or buds opposite each other at each node on a stem.

Orbicular Circular in outline (Plate 79, Leaf Form).

Ovary The part of the pistil that contains the ovules and later the seeds, and becomes a fruit (Plate 78, figures 3 and 4).

Ovate Egg-shaped in outline, broadest near the base (Plate 79, Leaf Form).

Ovoid An egg-shaped body.

Ovule Egg-containing body, which after fertilization becomes a seed.

Palmate Shaped like an open hand or palm. Radiately lobed, or with 3 or more leaflets arising from one point. A palmately veined leaf has 3 or more veins radiating from the junction with the petiole (Plate 79, Leaf Veins).

Panicle An irregularly branched, compound flower cluster (Plate 79, Leaf Veins).

Pappus The modified calyx limb in the family Compositae, forming a crown of bristles, awns or scales.

Pedicel The stalk of a single flower in a cluster (Plate 78, figure 3).

Peduncle The stalk of a flower cluster, or of a solitary flower (Plate 78, figure 3).

Pendulous Hanging down; drooping.

Perianth The outer part of a flower consisting of the calyx and corolla.

Perfect (flower) Bisexual: having functional stamens and a functional ovary.

Persistent Not falling off.

Petal A unit, or lobe, of the corolla.

Petiole The leaf stalk (Plate 78, figure 1).

Pilose Having soft, long, straight hairs.

Pinnate Having lateral appendages attached successively along a central axis. A pinnately compound leaf has leaflets attached along each side of the rachis. A pinnately veined leaf has lateral veins departing successively on either side of the midvein (Plate 79, Leaf Veins).

Pistil The ovule-bearing and seed-bearing organ in a flower, composed of 1 carpel or of 2 or more connate carpels (Plate 78, figure 4).

Pistillate A flower that has a pistil but no stamens. A unisexual plant bearing pistillate (female) flowers only.

Pome An fleshy applelike fruit, formed from an inferior ovary and its adherent flower base, the seeds in thin-walled chambers.

Precocious Flowering before emergence of the leaves.

Pruinose Covered with a bluish powdery bloom that is easily rubbed off.

Puberulent Minutely pubescent, the hairs barely visible.

Pubescent Covered with short, soft hairs.

Pungent Having a sharp, stiff point.

Raceme Inflorescence with stalked flowers on equal-lengthed pedicels, attached successively along a cental axis (Plate 79, Inflorescences).

Rachis The central axis of a raceme or spike of flowers, or of a compound leaf (Plate 78, figure 1).

Receptacle Specialized stem tip bearing some or all of the flower parts.

Recurved Curved downward or backward.

Reticulate Net-veined, as opposed to parallel-veined.

Revolute (leaf margin) Rolled or curved under, toward the lower or dorsal side.

Rhizome A perennial subterranean stem with buds, often running horizontally. Rhizomes enable a plant to form a spreading colony vegetatively.

Samara Fruit with usually one seed and a wing or wings for dispersal by wind.

Scales (1) In flowering plants, modified small leaves, such as bud scales or bracts (Plate 77, figure 5; Plate 78, figure 1). (2) In conifers, modified branches, forming the overlapping seed-bearing cone scales (Plate 77, figures 1 and 2). (3) In Elaeagnaceae, modified epidermal hairs, forming a scurfy surface.

Scurfy Covered with loose, thin scales.

Sepal A division of the calyx: an outer appendage of the perianth (Plate 78, figures 3 and 4).

Septicidal Referring to a capsule that opens along the partitions.

Sericeous Covered with parallel, shiny hairs that give the surface a silky or satiny lustre.

Serotinous Flowering in summer, after leaves have expanded.

Serrate With uniform oblique teeth, as on a saw (Plate 79, Leaf Margin).

Serrulate Finely serrate.

Sessile Not stalked.

Simple (simply) Having only one part or being one size. A simple leaf has only one blade. A simply serrate leaf has teeth of only one size on its margin.

Sinuate Having smoothly curved lobes and sinuses (Plate 79, Leaf Margin).

Sinus The recess between the lobes of a leaf (Plate 78, figure 1).

Spatulate Narrowed downward from a broad apex (Plate 79, Leaf Form).

Stamen The pollen-bearing organ of the flower (Plate 78, figure 3 and 4).

Staminate A flower that has stamens but no pistils. A unisexual plant bearing staminate (male) flowers only.

Stellate Starlike (of branching hairs).

Steppe Open prairie, dominated by grasses, sagebrush or other shrubs; usually treeless, but sometimes sparsely treed, as with Ponderosa Pine.

Stigma The tip of a pistil (Plate 78, figures 3 and 4). Its surface receives the pollen.

Stipe The stalklike support of an ovary.

Stipule An appendage at the base of a petiole (Plate 78, figure 1).

Stomata (singular: stoma) Small pores in the surface of a leaf for diffusion of air into the plant. In conifers, stomata usually appear as fine bluish-white lines, or dots in longitudinal rows.

Style The part of the pistil between the ovary and the stigma (Plate 78, figures 3 and 4).

Subcordate Almost but not quite heart-shaped.

Subglobose Nearly globose. (See Depressed-globose.)

Subshrub A barely shrubby, almost herbaceous plant, low growing and often woody at the base with herbaceous shoots.

Subtend Stand below, so as to enclose or surround, as a bract beneath a flower: the bract subtends the flower in its axil.

Superior (ovary) Borne above and free from the bases of the outer floral parts.

Suture A line where organs or their parts are joined together, or where they will split open.

Tomentose Clothed with thick, matted hairs.

Trifoliolate Having three leaflets. (See also Foliolate.)

Truncate Ending abruptly, as if cut off (Plate 79, Leaf Base).

Two-ranked (leaves) Diverging from the twig alternately on opposite sides, as in Elms.

Umbel An inflorescence with pedicels arising from a single point (Plate 79, Inflorescences).

Umbellet A secondary or small umbel.

Valve A part of a longitudinally dividing fruit.

Villous Covered with long, soft, usually curved or curly hairs; shaggy.

Whorl A ring of leaves or other organs around a single node.

Abbreviations and Symbols

B.S.P. Britton, N.L., E.E. Sterns & J.F. Poggenburg.

DC. Candolle, Augustin Pyramus de.

L. Linnaeus, Carolus (Carl von Linné) the elder.

L.f. Linné, Carl von, the younger (son of L.).

lat. latitude.

long. longitude.

subg. subgenus.

subsp. subspecies.

var. variety.

× hybrid. When prefixed to a specific name, indicates that it is a named hybrid. When placed between specific names, indicates parentage of a hybrid, as a formula; e.g., *Pinus* × *murraybanksiana* is the named hybrid progeny resulting from the cross *Pinus banksiana* × *contorta*.

REFERENCES

1   Argus, G.W. 1973. *The Genus* Salix *in Alaska and Yukon.* Publications in Botany, no. 2. Ottawa: National Museums of Canada.

2   Boivin, B. 1959. *Abies balsamea* (Linne) Miller *et ses variations. Le Naturaliste Canadien* 86: 220-223.

3   Brayshaw, T.C. 1960. *Key to the Native Trees of Canada.* Bulletin 125. Ottawa: Canada Department of Forestry.

4   Brayshaw, T.C. 1970. The dry forests of southern British Columbia. *Syesis* 3: 17-43.

5   Brayshaw, T.C. 1976. *Catkin Bearing Plants of British Columbia.* Occasional Paper no. 18. Victoria: British Columbia Provincial Museum.

6   Brayshaw, T.C. 1989. *Buttercups, Waterlillies and Their Relatives (the Order Ranales) in British Columbia.* Memoir no. 1. Victoria: Royal British Columbia Museum.

7   Brayshaw, T.C. 1996. *Catkin-Bearing Plants of British Columbia.* 2nd rev. ed. Victoria: Royal British Columbia Museum.

8   Brunsfeld, S.J., and F.D. Johnson. 1990. Cytological, morphological, ecological and phenological support for specific status of *Crataegus suksdorfii* (Rosaceae). *Madrono* 37(4): 274-282.

9   Calder, J.A., and R.L. Taylor. 1968. *Flora of the Queen Charlotte Islands: Part 1.* Monograph no. 4, Part 1. Ottawa: Research Branch, Department of Agriculture Canada.

10 Clapham, A.R., T.G. Tutin and E.F. Warburg. 1958. *Flora of the British Isles.* Cambridge: Cambridge University Press.

11 Cronquist, Arthur. 1979. *How to Know the Seed Plants.* Dubuque, Iowa: Wm C. Brown Co.

12 Cunningham, G.C. 1958. *Forest Flora of Canada.* Bulletin 121. Ottawa: Department of Forestry Canada.

13 Davidson, J. 1927. *Conifers, Junipers and Yew: Gymnosperms of British Columbia.* London: T. Fisher Unwin.

14 Dawson, G.M. 1881. *Report on an Exploration from Port Simpson on the Pacific Coast to Edmonton on the Saskatchewan: A Portion of the Northern Part of British Columbia and the Peace River Country, 1879.* Report of Progress, 1879-80, Pt. B. Montreal: Geologic and Natural History Survey of Canada.

15 Dorn, R.D. 1975. A systematic study of *Salix* section *cordatae* in North America. *Canadian Journal of Botany* 53(15):1491-1522.

16 Douglas, G.W., G.B. Straley and D. Meidinger. 1989-1994. *The Vascular Plants of British Columbia.* 4 vols. Victoria: B.C. Ministry of Forests.

17 Douglas, S. 1991. *Trees and Shrubs of the Queen Charlotte Islands.* Queen Charlotte City, B.C.: Islands Ecological Research.

18 Dugle, J.R. 1966. A taxonomic study of western Canadian species in the genus *Betula. Canadian Journal of Botany* 44: 929-1007.

19 Eastham, J.W., 1947. *Supplement to Flora of Southern British Columbia (J.K. Henry).* Special Publication no. 1. Victoria: British Columbia Provincial Museum.

20 Flora of North America Editorial Committee. 1993. *Flora of North America,* vol. 2: *Pteridophytes and Gymnosperms.* New York: Oxford University Press.

21 Franklin, J.F. 1961. *A Guide to Seedling Identification for 25 Conifers of the Pacific Northwest.* Portland: Pacific Northwest Forest and Range Experiment Station, Forest Service, U.S. Department of Agriculture.

22 Garman, E.H. 1957. *The Occurrence of Spruce in the Interior of British Columbia.* Publication T, 49. Victoria: British Columbia Forest Service.

23 Garman, E.H. 1973. *Guide to the Trees and Shrubs of British Columbia.* Handbook no. 31. Victoria: British Columbia Provincial Museum.

24 Glendenning, R.G. 1934. Notes on the distribution of Garry Oak in British Columbia. *Forestry Chronicle* 10:207-208.

25 Glendenning, R.G. 1948. Occurrence of a columnar form of the Western Red Cedar. *Canadian Field-Naturalist* 62:39-40.

26 Greuter, W., ed. 1988. *International Code of Botanical Nomenclature.* Regnum vegetabile, vol. 118. Königstein, Germany: Koeltz Scientific Books.

27 Harlow, W.M., and E.S. Harrar. 1958. *Textbook of Dendrology.* 4th ed. New York: McGraw-Hill.

28 Henry, J.K. 1915. *Flora of Southern British Columbia and Vancouver Island.* Toronto: W.J. Gage.

29 Hitchcock, C.L., A. Cronquist, M. Ownbey and J.W. Thompson. 1955-1966. *Vascular Plants of the Pacific Northwest.* In 5 parts: Part 1, 1969; Part 2, 1964; Part 3, 1961; Part 4, 1959; Part 5, 1955. Seattle: University of Washington Press.

30 Hitchcock, C.L., and A. Cronquist. 1973. *Flora of the Pacific Northwest.* Seattle: University of Washington Press.

31 Hosie, R.C. 1979. *Native Trees of Canada.* 8th ed. Don Mills, Ontario: Fitzhenry and Whiteside.

32 Hulten, E. 1967. Comments on the flora of Alaska and Yukon. *Arkiv för Botanik* serie 2 band 7 nr. 1:99-100.

33 Hulten, E. 1968. *Flora of Alaska and Neighboring Territories.* Stanford, California: Stanford University Press.

34 Krajina, V.J. 1954. *Second Annual Report on Ecological Classification of Hemlock Forests, Columbia River Basin.* Mimeograph. Vancouver.

35 Krajina, V.J. 1959. *Biogeoclimatic Zones in British Columbia.* Botanical Series no. 1. Vancouver: University of British Columbia.

36 Krussmann, G. 1984-1986. *Manual of Cultivated Broad-leafed Trees and Shrubs.* Translated 1984 by Michael Epp. Technical Editor: Gilbert S. Daniels. Portland: Timber Press.

37 Krussmann, G. 1985. *Manual of Cultivated Conifers.* 2nd ed. Edited by Hans-Dieter Warda. Translated by Michael Epp. Technical Editor: Gilbert S. Daniels. Portland: Timber Press.

38 Laidlaw, W.B.R. 1960. *Guide to British Hardwoods.* London: Leonard Hill.

39 Little, E.L., Jr. 1971. *Atlas of United States Trees,* vol. 1: *Conifers and Important Hardwoods.* Misc. Publication no. 1146. Washington, D.C.: Forest Service, U.S. Department of Agriculture.

40 Little, E.L., Jr. 1976. *Atlas of United States Trees* vol. 3: *Minor Western Hardwoods.* Misc. Publication no. 1314. Washington, D.C.: Forest Service, U.S. Department of Agriculture.

41 Little, E.L., Jr. 1979. *Check List of United States Trees (Native and Naturalized).* Agriculture Handbook no. 541. Washington, D.C.: Forest Service, U.S. Department of Agriculture.

42 Love, R., and M. Feigen. 1978. Interspecific hybridization between native and naturalized *Crataegus* (Rosaceae) in western Oregon. *Madrono* 25: 211-217.

43 Lyons, C.P. 1952. *Trees, Shrubs and Flowers to Know in British Columbia.* Vancouver: J.M. Dent.

44 MacKinnon, A., J. Pojar and R. Coupé, eds. 1992. *Plants of Northern British Columbia.* Victoria: British Columbia Forest Service; Edmonton: Lone Pine Publishing.

45 Meidinger, D. 1987 (rev. 1988). *Recommended Vernacular Names for Common Plants of British Columbia.* Internal report. Victoria: Research Branch, British Columbia Ministry of Forests and Lands.

46 Moss, E.H. 1983. *Flora of Alberta.* 2nd ed., revised by J.G. Packer. Toronto: University of Toronto Press.

47 Ogilvie, R.T. 1989. Distribution and ecology of Whitebark Pine in western Canada. In *Proceedings, Symposium on Whitebark Pine Ecosystems: Ecology and Management of High-altitude Resources.* General Technical Report INT-270: 54-60. Ogden, UT: Forest Service, U.S. Department of Agriculture.

48 Popova, T.N. 1972. *Vaccinium.* In *Flora Europaea,* vol. 3, edited by T.G. Tutin, V.H. Heywood, N.A. Burgess, D.M. Moore, D.H. Valentine, S.M. Walters and D.A. Webb. Cambridge: Cambridge University Press.

49 Raup, H.M. 1934. *Phytogeographic Studies in the Peace and Upper Liard River Regions, Canada.* Contributions from the Arnold Arboretum of Harvard University, Jamaica Plains, Mass., 6:2-230.

50 Raup, H.M. 1947. The botany of southwest Mackenzie. *Sargentia* 6:1-275.

51 Raup, H.M. 1959. *The Willows of Boreal Western America.* Contributions from the Gray Herbarium of Harvard University, Jamaica Plains, Mass., 185:3-95.

52 Rehder, A. 1949. *Manual of Cultivated Trees and Shrubs Hardy in North America.* 2nd ed. New York: Macmillan.

53 Rowe, J.S. 1972. *Forest Regions of Canada.* Publication no. 1300. Ottawa: Canadian Forestry Service.

54 Schmidt, R.L. 1957. *The Silvics and Plant Geography of the Genus Abies in the Coastal Forests of British Columbia.* Technical Publication T. 46. Victoria: British Columbia Forest Service, Department of Lands and Forests.

55 Stoltmann, R. 1987. *Hiking Guide to the Big Trees of Southwestern British Columbia.* Western Wilderness Committee.

56 Szczawinski, A.F. 1962. *The Heather Family (Ericaceae) of British Columbia.* Handbook no. 19. Victoria: British Columbia Provincial Museum.

57 Taylor, T.M.C. 1959. The taxonomic relationship between *Picea glauca* (Moench) Voss and *P. engelmannii* Parry. *Madrono* 15(4):111-115.

58 Taylor, T.M.C. 1973. *The Rose Family (Rosaceae) of British Columbia.* Handbook no. 30. Victoria: British Columbia Provincial Museum.

59 Taylor, T.M.C. 1974. *The Pea Family (Leguminosae) of British Columbia.* Handbook no. 32. Victoria: British Columbia Provincial Museum.

60 Taylor, T.M.C. 1974. *The Figwort Family (Scrophulariaceae) of British Columbia.* Handbook no. 33. Victoria: British Columbia Provincial Museum.

61 Tucker, J.M., and J.R. Maze. 1973. The Revelstoke oaks. *Syesis* 6:41-46.

62 Vander Kloet, S.P. 1988. *The Genus* Vaccinium *in North America.* Publication no. 1828. Ottawa: Research Branch, Agriculture Canada.

63 Viereck, L.A., and E.L. Little, Jr. 1972. *Alaska Trees and Shrubs.* Publication no. 410. Washington, D.C.: Forest Service, U.S. Department of Agriculture.

INDEX

Trees and Shrubs of British Columbia,
written and illustrated by T. Christopher Brayshaw.

Edited, designed and typeset in Times Roman 10/12
by Gerry Truscott, RBCM.

Copy edited by Frank Chow.

All photographs by T. Christopher Brayshaw, except for the cover
photographs of *Rosa nutkana* (front) and *Arbutus menziesii* (back) by
Richard Hebda, RBCM.

Cover and colour section design by Chris Tyrrell, RBCM.